T0007787

HOT MOLECULES, COLD ELECTRONS

FIGURE A. Father Neptune Blesses Uncle Sam and Britannia on the Success of the 1866 Trans-Atlantic Telegraph Cable. Reproduced from the August 11, 1866, issue of *Punch*.

OUR TWO HEROES

FIGURE B. Joseph Fourier (1768–1830) of France. Courtesy of the Deutsches Museum, Munich.

FIGURE C. William Thomson (1824–1907) of Scotland. Courtesy of the Institution of Electrical Engineers, London.

HOT MOLECULES, COLD ELECTRONS

From the Mathematics of Heat to the Development of the Trans-Atlantic Telegraph Cable

PAUL J. NAHIN

PRINCETON UNIVERSITY PRESS

Princeton and Oxford

Copyright © 2020 by Princeton University Press

Requests for permission to reproduce material from this work
should be sent to permissions@press.princeton.edu

Published by Princeton University Press
41 William Street, Princeton, New Jersey 08540
99 Banbury Road, Oxford OX2 6JX

press.princeton.edu

All Rights Reserved

LCCN 2019027369
First paperback printing, 2022
Paper ISBN 9780691207841
Cloth ISBN 9780691191720
ISBN (ebook) 9780691199948

British Library Cataloging-in-Publication Data is available

Editorial: Susannah Shoemaker and Lauren Bucca
Production Editorial: Debbie Tegarden
Production: Jacquie Poirier
Publicity: Matthew Taylor and Katie Lewis
Copyeditor: Cyd Westmoreland

Jacket /Cover images: Transatlantic cable design templates, 1850–1860s. Science History
Images / Alamy Stock Photo. Other cross-sections: ibusca, duncan1890 / iStock

This book has been composed in STIX fonts and Futura

"We have not written for physicists or for beginners."

—from the Preface of the 1943 classic *Fourier Series* by Hardy and Rogosinski, a sentiment *not* embraced by the author of this book

"If only I had more mathematics."

—Einstein, the day before he died, illustrating that you can *never* know too much math

fortunately, however

"In this book there is no mathematics that exceeds that of a college freshman."

—the author's promise, so if you love math, you will almost certainly be able to read this entire book (with only a few possible exceptions here and there)[1]

And once you've done that you'll be able to derive and solve the heat equation

$$\frac{\partial u}{\partial t} = k \frac{\partial^2 u}{\partial x^2},$$

which explains how milk and sugar dissolve in your coffee, how the temperature changes in a molten planet cooling down to become one with a solid crust (the origin of the Earth), and how electricity "traveled" through the Atlantic cable.

---

1 Those exceptions, however, *can* be read and understood by any college freshman who applies herself, and who will then be an exceptionally well-read freshman and so the recipient of boundless admiration from all who know her.

To the memory of

Barbara Jean Beechler (1928–2003)

Late professor emerita of mathematics, Pitzer College,
Claremont, California

Barbara was a pure mathematician who nonetheless saw nothing strange in an engineer penning math books, and so enthusiastically encouraged me to continue writing. If, in some magical way, she could somehow read this book, I hope she would be pleased.

An engineer thinks that his equations are an approximation to reality.
A physicist thinks reality is an approximation to his equations.
A mathematician doesn't care.

—An anonymous observer who was either a deep thinker or a cynic (or even, perhaps, both). After reading this book, perhaps you'll have your own opinion.

It is positively spooky how the physicist finds the mathematician
has been there before him or her.

—Steven Weinberg, 1979 Nobel laureate in physics, at a 1986 math conference

CONTENTS

5. William Thomson and the Infinitely Long Telegraph Cable Equation

6. Epilogue

FOREWORD

Fascinating questions and engaging narrative, together with significant yet graspable mathematics, have long been Paul Nahin's way. This approach is exemplified in *Hot Molecules, Cold Electrons*, whose title strikingly links the theory of heat flow and the physics of the transatlantic cable. But Nahin links the topics much more profoundly. Assuming no more than a year of university-level calculus, he derives Fourier's equation for how heat flows and applies it to describe things as diverse as how milk and sugar dissolve in coffee; how the temperature changes in a molten Earth cooled down to become the one with a solid crust we have now; and, in a final tour de force, how electricity "travels" through the Atlantic cable.

Who is the audience for this book? Every engineer or scientist who wants to look under the hood of the standard results and conclusions, and every teacher who wants help in motivating the science and mathematics. Nahin doesn't assume a course in Fourier analysis or even experience with trigonometric series. A year of university-level calculus—provided, of course, that the reader is willing to make the effort to follow the arguments—is enough for Nahin's exposition. As he says, this mathematical approach he presents, "in addition to being historical . . . is just astonishingly clever and not to be missed." So the audience also should include students who wonder what the calculus is really for.

"The calculus was the first achievement of modern mathematics," John von Neumann once said, "and it is difficult to overestimate its importance." Yet the modern calculus student may finish the course with the impression that the purpose of the calculus is to find tangents to polynomial curves, volumes of odd shapes, or the velocity of the bottom of a sliding ladder. Such impressions will give way to enthusiastic agreement with von Neumann when the student reads Nahin's book. The more mathematically adept reader will also enjoy the presentation, first wondering and then marveling at the way Nahin deploys relatively elementary methods to explain the heat equation and its solution, as well as a wealth of other useful results in real and complex analysis. He uses modern software to produce instructive graphs, and gives computer-generated numerical calculations to test symbolically derived results. Thus, he shows the reader how applied mathematics is actually done. And Nahin levels with us about what he's doing and why, saying that all this mathematics isn't in the book "because I want to make your life difficult—*it's here because that's the way the world is made.*"

Nahin highlights a crucial, though counterintuitive, fact about the way the world of the history of mathematics is made. A piece of mathematics often is developed to solve one set of problems, and then, later on, scientists find that this is exactly the mathematics needed to solve a quite different problem in the natural world. Nahin gives the example of the 19th-century mathematicians who "developed the math needed by Einstein long before Einstein knew he would need it." But the most important case for this book is the 18th-century mathematics that let Fourier develop his theory of heat.

The story Nahin tells about Fourier's mathematical models of heat flow presents the historical background in a lively way, which helps the reader understand the mathematical part of the story. When he introduces a new term, he explains why that particular word is used. When he introduces the name of a mathematician, he explains who that mathematician is. He highlights especially beautiful arguments and encourages the reader to share his delight in them.

Equally delightful is a fair amount of material that isn't mathematics that a mathematical reader may want to read aloud to friends and family. For instance, charming footnotes support the overall story, like the one describing how the Comte de Buffon used his bare hands to evaluate the temperature of cooling iron spheres. There are also many apt quotations, like Lord Kelvin's saying, "The age of the earth as an abode fitted for life . . . [is] as important as the date of the Battle of Hastings . . . for English History." Computer scientists may enjoy Nahin's description of how Kelvin imagined an analog device that could evaluate the partial sums of the Fourier series used to predict the timing of tides by using analogous mechanical forces, and Nahin's use of modern software to answer questions about the transmission of information via the Atlantic cable.

I especially enjoyed his observation that submarine cables, which still exist today, are hard to hack into.

Similar engaging material introduces the Atlantic cable of 1866, before Nahin turns to the science involved. He first invites us along by saying, "Off we go to the bottom of the sea." The Atlantic cable was 2,000 miles long and ran 3 miles beneath the surface of the ocean. It enabled real-time communication at a rate of 6–8 words per minute between the Old World and the New for the first time. Nahin describes the social context, including the Victorian novels by Jules Verne and H. G. Wells, and cites the objection that transatlantic communication would waste money by communicating many things that were trivial. Nahin describes how, regardless of objections, Cyrus Field helped raise 1.5 million dollars for the project, quite a lot of money in the middle of the 19th century. Thus, Nahin says, "the Atlantic cable was a techno-logical masterpiece, the result of a flawless amalgamation of science and engineering, combined with real-world capitalism and politics."

And then Nahin returns to the mathematics, which provides its own kind of excitement as the key connecting idea is explained: The flow of electricity in a semi-infinite induction-free telegraph cable is described by Fourier's heat equation (with appropriate choice of constants), and that, therefore, "all of Fourier's results for heat flow in a one-dimensional, semi-infinite rod, with its lateral surface insulated" carry over to that electrical flow. And then he shows us how all this is done.

The book is fittingly dedicated to my late colleague and beloved friend, Barbara Beechler, who founded the mathematics program at Pitzer Col-lege, and always championed the power of mathematics and the impor-tance of explaining it superbly. I imagine her now, waving this book at colleagues and students, telling everyone, "*This* is how it should be done!" *Hot Molecules, Cold Electrons* exemplifies the way mathematics can both reveal the structure of the natural world and enable discoveries that transform human existence.

Judith V. Grabiner
Flora Sanborn Pitzer Professor of Mathematics Emerita
Pitzer College, Claremont, California
December 24, 2018

HOT MOLECULES, COLD ELECTRONS

CHAPTER 1
Mathematics and Physics

1.1 Introduction

When Isaac Newton showed the intimate connection between celestial mechanics and math by extending the inverse-square law of gravity from mere earthly confines to the entire universe, and when James Clerk Maxwell used math to join together the erstwhile separate subjects of magnetism and electricity, they gave science two examples of the *mutual embrace* (to use Maxwell's words) of math and physics. They had performed what are today called the first two *unifications* of mathematical physics.

Two centuries separated those two unifications, but the next one came much faster, with Albert Einstein's connection of space and time in the special theory of relativity and then, soon after, together with gravity in the general theory, less than a century after Maxwell. Again, it was mathematics that was the glue that sealed the union, but now there was a significant difference. With Newton and Maxwell, the required math was already known to physicists beforehand; but with Einstein, it was not. Einstein was an excellent *applied* mathematician, but he was not a creator of new math and so, in the early 1900s, he was in a semi-desperate state.

As Einstein himself put it, "I didn't become a mathematician because mathematics was so full of beautiful and difficult problems that one might waste one's power in pursuing them without finding the central problem."[1] When he needed tensor calculus to codify the physical principles of general relativity, he had to plead for aid from an old friend, a former fellow student who had helped him pass his college math exams.[2] As one of Einstein's recent biographers has memorialized this interesting situation, when he realized he didn't have the necessary math to express his insights into the *physics* of gravity, Einstein exclaimed "Grossman, you've got to help me or I will go crazy."[3] And, good friend that he was, Einstein's pal Marcel *did* help. And that's how Einstein learned how to mathematically express what he knew *physically*, and thus were born the beautiful, coupled, nonlinear partial differential equations of general relativity that generations of theoretical physicists have wrestled with now for over a century.

I tell you all this for two reasons. First, it's *not* to tease the memory of Einstein (who was, I surely don't have to tell you, a once-in-a-century genius), but rather to heap praise on the mathematicians—people like the German Bernhard Riemann (1826–1866) and the Italians Gregorio Ricci-Curbastro (1853–1925) and Tullio Levi-Civita (1873–1941)—who had developed the math needed by Einstein long before Einstein knew he would need it. And second, because there is an earlier, equally dramatic but not so well-known occurrence of this sort of anticipatory good fortune in mathematical physics. It is that earlier story that has inspired this book.

1.2 Fourier and *The Analytical Theory of Heat*

Sometime around 1804 or so (perhaps even as early as 1801), the French mathematical physicist and political activist Jean Baptiste Joseph Fourier (1768–1830)—who came perilously close to being separated from his world-class brain by the guillotine during the Terror of the French Revolution[4]—began his studies on how heat energy propagates in solid matter. In other words, it was then that he started pondering the physics of *hot molecules in bulk* (and so now you can see where the first half of the title of this book comes from). In the opening of his masterpiece, *The Analytical Theory of Heat* (1822), about which I'll say more in just a bit, Fourier tells us *why* he decided to study heat: "Heat, like gravity, penetrates every substance of the universe. . . . The object of our work is to set forth the mathematical laws [of heat]. The theory of heat will hereafter form one of the most important branches of general physics."

A few years after beginning his studies (1807), he had progressed far enough to write a long report of his work called *On the Propagation of Heat in Solid Bodies*, which received some pretty severe criticism. The critics weren't quacks, but rather included such scientific luminaries as Joseph-Louis Lagrange (1736–1813) and Pierre-Simon Laplace

(1749–1827), who, while certainly pretty smart fellows themselves, nonetheless stumbled over the sheer novelty of Fourier's math. Fourier, you see, didn't hesitate to expand arbitrary periodic functions of space and time in the form of infinite sums of trigonometric terms (what we today call *Fourier series*). Lagrange and Laplace just didn't think that was possible. Fourier, of course, was greatly disappointed by the skepticism. But fortunately, he was not discouraged by the initial lack of enthusiasm. He didn't give up, and continued his studies of heat in matter.[5]

In 1817 Fourier's talent was formally recognized, and he was elected to the French Academy of Sciences, becoming in 1822 the secretary to the mathematical section. That same year finally saw the official publication of his work on heat, a work that is still an impressive read today. In *The Analytical Theory of Heat*, Fourier included his unpublished 1807 effort, plus much more on the representation of periodic functions as infinite sums of trigonometric terms. His *mathematical* discoveries on how to write such series were crucial in his additional discoveries on how to solve the fundamental *physics* equation of heat flow, the aptly named *heat equation*, which is (just to be precise) a second-order partial differential equation. (This will prove to be *not* so scary as it might initially sound.)

In the following chapters of the first part of this book, we'll develop Fourier's math, then derive the heat equation from first principles (conservation of energy), and then use Fourier's math to solve the heat equation and to numerically evaluate some interesting special cases (including a calculation of the age of the Earth). Then, in the penultimate chapter of the book, I'll show you how the man who did that calculation of the age of the Earth—the Irish-born Scottish mathematical physicist and engineer William Thomson (1824–1907)—discovered a quarter-century after Fourier's death that the heat equation is also the defining physics, under certain circumstances, of a very long submarine telegraph cable (in particular, the famous trans-Atlantic electric cables of the mid-19th century).

Thomson, who was knighted by Queen Victoria in 1866 for his cable work (and later, in 1892, was elevated to the peerage to become the famous Lord Kelvin), directly used and credited Fourier's mathematics in his pioneering study of electric communication cables. The Atlantic cables, in particular, lay deep (in places, up to 15,000 feet beneath the surface) in the cold waters of the Atlantic. And since electric current is caused by the motion of electrons, you now see where the second half of the title of this book comes from.

Telegraphy was the very first commercial application of electricity, being introduced in England by railroad operators as early as 1837. This date is doubly impressive when it is remembered that the electric battery (*voltaic pile*) had been invented by the Italian Alessandro Volta (1745–1827) less than 40 years before. Then, less than 70 years after the battery, messages were being routinely sent through a submarine cable thousands of miles long lying nearly 3 miles beneath the stormy Atlantic Ocean, an accomplishment that struck the imaginations of all but the dullest like a thunderbolt—*it was nothing less than a miracle*—and the men behind the creation of the trans-Atlantic cable became scientific and engineering superstars. What you'll read in this book is the mathematical physics of what those men did, based on the mathematical physics of Fourier's theory of the flow of heat energy in matter.

The technical problems discussed in this book are routinely attacked today by electrical engineers using a powerful mathematical technique generally called the *operational calculus* (specifically, the *Laplace transform*). The transform had been around in mathematics long before the engineers became aware of it in the 1930s, but it was *not* the tool Fourier and Thomson used to solve the equations they encountered. They instead used the classical mathematical techniques of their day, what is called *time domain* mathematics, rather than the modern transform domain approach of engineers. Fourier and Thomson were enormously clever analysts, and since my intention in this book is to weave the historical with the technical, everything you read here is just how either man might have written this book. There are *lots* of other books available that discuss the transform approach, and I'll let you look one up if you're curious.[6]

Now, before we do anything else, let me first show you a little math exercise that is embedded, with little fanfare, in *The Analytical Theory of Heat*, one that uses nothing but high school AP-calculus. It illustrates how an ingenious mind can extract amazing results from what, to less clever minds, appears to be only routine, everyday stuff. What I am about to show you is a mere half page in *Analytical Theory*, but I'm going to elaborate (that is, inflate) it all just a bit to make sure I cover all bets. My reference is the classic 1878 English translation from the original French by Alexander Freeman (1838–1897), who was a Fellow at St. John's College, Cambridge. (You can find exactly what Fourier wrote on page 153 of the 1955 Dover Publications reprint—itself now a minor classic—of Freeman's translation.) Going through this preliminary exercise will sharpen your appreciation for the genius of Fourier.

1.3 A First Peek into Fourier's Mathematical Mind

We'll start with something you could just look up in a math handbook, but, since I want to impress you with how a good high school student (certainly a college freshman) could do all that follows with nothing but a stick to write with on a sandy beach, let's begin by deriving the indefinite integration formula

$$\int \frac{dx}{1+x^2} = \tan^{-1}(x) + C,$$

where C is an arbitrary constant.

Look at Figure 1.3.1, which shows a right triangle with perpendicular sides 1 and x, and so geometry (the Pythagorean theorem) says the hypotenuse is $\sqrt{1+x^2}$. The base angle is denoted by θ, and so we have, by construction,

(1.3.1) $$x = \tan(\theta).$$

If we differentiate (1.3.1) with respect to x, we'll get

$$1 = \frac{d}{dx}\tan(\theta) = \frac{d}{dx}\left\{\frac{\sin(\theta)}{\cos(\theta)}\right\} = \frac{\cos^2(\theta)\dfrac{d\theta}{dx} + \sin^2(\theta)\dfrac{d\theta}{dx}}{\cos^2(\theta)}$$

or,

$$1 = \frac{d\theta}{dx}\frac{\cos^2(\theta) + \sin^2(\theta)}{\cos^2(\theta)}$$

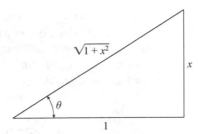

FIGURE 1.3.1. A right triangle, with x any positive value.

or, recalling the identity $\cos^2(\theta) + \sin^2(\theta) = 1$,

$$1 = \frac{d\theta}{dx} \frac{1}{\left\{ \dfrac{1}{\sqrt{1+x^2}} \right\}^2} = \frac{d\theta}{dx}(1+x^2).$$

That is,

$$(1.3.2) \qquad \frac{d\theta}{dx} = \frac{1}{1+x^2}.$$

But from (1.3.1), we have

$$\theta = \tan^{-1}(x)$$

and so, putting that into (1.3.2), we have

$$\frac{d}{dx}\tan^{-1}(x) = \frac{1}{1+x^2},$$

which, when integrated indefinitely, instantly gives us our result:

$$(1.3.3) \qquad \tan^{-1}(x) + C = \int \frac{dx}{1+x^2},$$

where C is some (any) constant.

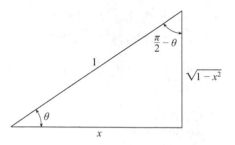

FIGURE 1.3.2. Another right triangle.

Okay, put (1.3.3) aside for now, and look at Figure 1.3.2, which shows a right triangle with a hypotenuse of 1, and one of the perpendicular sides as x. Thus, the other perpendicular side is $\sqrt{1-x^2}$. Since Euclid tells us that the two acute angles sum to $\frac{\pi}{2}$ radians, we can immediately write

(1.3.4) $$\frac{\pi}{2} = \tan^{-1}\left\{\frac{\sqrt{1-x^2}}{x}\right\} + \tan^{-1}\left\{\frac{x}{\sqrt{1-x^2}}\right\}.$$

If we write

$$u = \frac{\sqrt{1-x^2}}{x},$$

then (1.3.4) becomes the identity

(1.3.5) $$\frac{\pi}{2} = \tan^{-1}\{u\} + \tan^{-1}\left\{\frac{1}{u}\right\}.$$

Okay, put (1.3.5) aside for now, and turn your attention to the claim

(1.3.6) $$\frac{1}{1+u^2} = 1 - u^2 + u^4 - u^6 + \cdots.$$

Do you see where this comes from? If not, just multiply through by $1+u^2$, and watch how (1.3.6) reduces to $1=1$, which is pretty hard to deny! (Or, just do the long division on the left of (1.3.6) directly.) Now, integrate both sides of (1.3.6) which, recalling (1.3.3), says

$$\int \frac{du}{1+u^2} = \tan^{-1}(u) + C = u - \frac{1}{3}u^3 + \frac{1}{5}u^5 - \frac{1}{7}u^7 + \cdots.$$

When $u = 0$, it's clear that the infinite sum on the right is zero, and since $\tan^{-1}(0) = 0$, then $C = 0$. That is,

(1.3.7) $$\tan^{-1}(u) = u - \frac{1}{3}u^3 + \frac{1}{5}u^5 - \frac{1}{7}u^7 + \cdots.$$

In particular, if we set $u = 1$, then (1.3.7) reduces to the beautiful (if computationally useless way to compute π, because convergence is *extremely* slow)

$$\tan^{-1}(1) = \frac{\pi}{4} = 1 - \frac{1}{3} + \frac{1}{5} - \frac{1}{7} + \cdots,$$

a result discovered by other means in 1682 by the German mathematician Gottfried Wilhelm von Leibniz (1646–1716). It is interesting to note, in passing, the similar-looking series discovered in 1668 by the Danish-born mathematician Nicolaus Mercator (1620–1687):

$$\ln(2) = 1 - \frac{1}{2} + \frac{1}{3} - \frac{1}{4} + \frac{1}{5} - \cdots.$$

Since (1.3.7) is an identity in u, it remains true if we replace every u with $\frac{1}{u}$, and so

(1.3.8) $$\tan^{-1}\left(\frac{1}{u}\right) = \frac{1}{u} - \frac{1}{3}\left(\frac{1}{u}\right)^3 + \frac{1}{5}\left(\frac{1}{u}\right)^5 - \frac{1}{7}\left(\frac{1}{u}\right)^7 + \cdots.$$

Thus, from (1.3.5), (1.3.7), and (1.3.8), we have

$$\frac{\pi}{2} = \left(u + \frac{1}{u}\right) - \frac{1}{3}\left(u^3 + \frac{1}{u^3}\right) + \frac{1}{5}\left(u^5 + \frac{1}{u^5}\right) - \cdots.$$

Then, as Fourier wrote, "if we now write $e^{x\sqrt{-1}}[= e^{ix}$, in modern nota-tion, where $i = \sqrt{-1}$] instead of u ... we shall have"

$$\frac{\pi}{2} = (e^{ix} + e^{-ix}) - \frac{1}{3}(e^{i3x} + e^{-i3x}) + \frac{1}{5}(e^{i5x} + e^{-i5x}) - \cdots.$$

Using a famous identity (sometimes called a "fabulous formula")[7] due to the Swiss mathematician Leonhard Euler (1707–1783), who published it in 1748,

$$e^{i\theta} = \cos(\theta) + i\,\sin(\theta),$$

it follows that

$$e^{ix} + e^{-ix} = 2\cos(x), e^{i3x} + e^{-i3x} = 2\cos(3x), e^{i5x} + e^{-i5x} = 2\cos(5x), \ldots,$$

and so on. Using this, Fourier then immediately wrote

(1.3.9) $\quad \dfrac{\pi}{4} = \cos(x) - \dfrac{1}{3}\cos(3x) + \dfrac{1}{5}\cos(5x) - \dfrac{1}{7}\cos(7x) + \cdots,$

which is one of his famous infinite sums of trigonometric functions that Lagrange so objected to in 1807. This gives Leibniz's sum when $x = 0$, but now we see that Fourier has gone far beyond Leibniz, declaring that the sum is $\frac{\pi}{4} \approx 0.785$ for *lots* of other values of x as well. This is, I think you'll agree, a pretty remarkable claim!

With the invention of electronic computers and easy-to-use pro-gramming languages, it is a simple matter to experiment numeri-cally with (1.3.9), and Figure 1.3.3 shows what (1.3.9) looks like, where the right-hand-side of (1.3.9) has been calculated from the first 100 terms of the sum (that is, up to and including $\frac{1}{199}\cos(199x)$) for each of 20,000 values of x uniformly distributed over the interval $-10 < x < 10$.

Suppose a function $f(t)$ is written in the form of a power series. That is,

$$f(t) = c_0 + c_1 t + c_2 t^2 + \cdots + c_n t^n + \cdots.$$

It's a freshman calculus exercise to show that all the coefficients follow from the general rule

$$c_n = \frac{1}{n!}\left(\left.\frac{d^n f}{dt^n}\right|_{t=0}\right), \; n \geq 1,$$

that is, by taking successive derivatives of $f(t)$, and after each differentiation, setting $t = 0$. (The $n = 0$ case means, literally, *don't* differentiate, just set $t = 0$.) In this way, it is found, for example, that

$$\sin(t) = t - \frac{1}{3!}t^3 + \frac{1}{5!}t^5 - \cdots,$$

$$\cos(t) = 1 - \frac{1}{2!}t^2 + \frac{1}{4!}t^4 - \cdots,$$

$$e^t = 1 + t + \frac{1}{2!}t^2 + \frac{1}{3!}t^3 + \frac{1}{4!}t^4 + \frac{1}{5!}t^5 + \cdots.$$

Now, in the last series, set $t = ix$. Then

$$e^{ix} = 1 + ix + \frac{1}{2!}(ix)^2 + \frac{1}{3!}(ix)^3 + \frac{1}{4!}(ix)^4 + \frac{1}{5!}(ix)^5 + \cdots$$

$$= 1 + ix - \frac{1}{2!}x^2 - i\frac{1}{3!}x^3 + \frac{1}{4!}x^4 + i\frac{1}{5!}x^5 + \cdots$$

$$= \left(1 - \frac{1}{2!}x^2 + \frac{1}{4!}x^4 - \cdots\right) + i\left(x - \frac{1}{3!}x^3 + \frac{1}{5!}x^5 + \cdots\right)$$

$$= \cos(x) + i\sin(x).$$

This is an identity in x, and so continues to hold if we replace every x with a θ to give us our result: $e^{i\theta} = \cos(\theta) + i\sin(\theta)$.

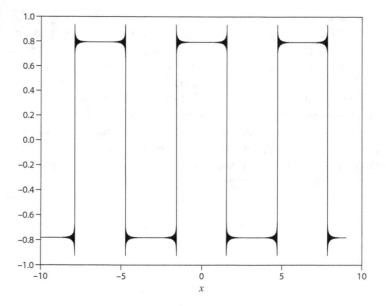

FIGURE 1.3.3. A computer-generated plot of Fourier's equation (1.3.9).

As you can see from Figure 1.3.3, the numerical "evidence" does appear to suggest that Fourier's math is correct—*sort of* (this numerical experiment means, admittedly, very little, if anything, to a pure mathematician, but it is quite compelling for physicists and engineers). The sum flips back and forth between two values,[8] $\frac{\pi}{4}$ and $-\frac{\pi}{4}$, with the swings occurring at odd multiples of $\frac{\pi}{2}$ for x. As Fourier states in *Analytical Theory* (page 144 in Freeman's translation), "the [sum] is $\frac{1}{2}\pi$ if [x] is included between 0 and $\frac{1}{2}\pi$, but . . . is $-\frac{1}{4}\pi$, if [x] is included between $\frac{1}{2}\pi$ and $\frac{3}{2}\pi$." Figure 1.3.3 also suggests that Fourier's comment implies that he did *not* actually make such a plot, because if he had, he would certainly have noticed all the dramatic oscillatory behavior around the transition points. In fact, Fourier made no comment at all on this hard-to-miss feature. For some reason, making such a plot wasn't done until 1848(!), when the twenty-two-year-old Henry Wilbraham (1825–1883) finally did so; his plots clearly show the oscillations.

After publishing his discovery in the *Cambridge and Dublin Mathematical Journal*, Wilbraham, a recent graduate of Trinity College, Cambridge, authored a few more mathematical papers and then, for some unknown reason, disappeared from the world of mathematics. When the oscillations were rediscovered again, many years later, they were named after somebody else: they are now called the *Gibbs phenomenon*, after the American mathematical physicist J. W. Gibbs (1839–1903), who briefly commented on them in an 1899 letter to the British science journal *Nature*. The oscillations occur in any Fourier series that represents a *discontinuous* function.[9] Mathematicians have known since 1906 that such a Fourier series converges to the *average* of the function's values on each side of the discontinuity when the series is evaluated *at* the point of discontinuity.

A dramatic calculation, one that also appears in *Analytical Theory*, is based on the integration of (1.3.9), which results in

$$(1.3.10) \quad \frac{\pi}{4}x = \sin(x) - \frac{1}{3^2}\sin(3x) + \frac{1}{5^2}\sin(5x) - \frac{1}{7^2}\sin(7x) + \cdots,$$

where the arbitrary constant of integration is zero (do you see why?—Evaluate (1.3.10) for $x=0$). (Figure 1.3.4, a plot of the right-hand side of (1.3.10), shows that while (1.3.9) is discontinuous, its integral is *continuous*.) If we substitute $x = \frac{\pi}{2}$ in (1.3.10), we get

$$(1.3.11) \quad \frac{\pi^2}{8} = 1 + \frac{1}{3^2} + \frac{1}{5^2} + \frac{1}{7^2} + \cdots.$$

Now, a quarter century before Fourier's birth, one of the great unsolved problems that had been taunting mathematicians for *centuries* was the calculation of

$$S = 1 + \frac{1}{2^2} + \frac{1}{3^2} + \frac{1}{4^2} + \frac{1}{5^2} + \cdots.$$

The calculation of S had quickly become the next obvious problem to attack after the surprising discovery by the 14th-century French math-

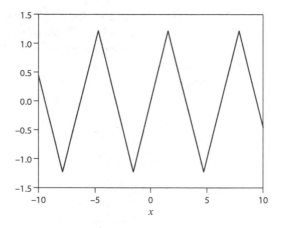

FIGURE 1.3.4. Integral of the right-hand side of (1.3.9).

ematician and philosopher Nicoleo Oresme (1320–1382) that the sum (called the *harmonic series*), defined as

$$H = 1 + \frac{1}{2} + \frac{1}{3} + \frac{1}{4} + \frac{1}{5} + \cdots,$$

diverges. This is a counterintuitive result for most people, because the terms seem to add *so very slowly* (for the partial sum to exceed 15 takes more than 1.6 million terms, and to reach a partial sum of 100 takes more than 1.5×10^{43} terms).

To prove that $H = \infty$ is not hard. Simply write H as

$$H = 1 + \frac{1}{2} + \left(\frac{1}{3} + \frac{1}{4} \right) + \left(\frac{1}{5} + \frac{1}{6} + \frac{1}{7} + \frac{1}{8} \right)$$
$$+ \text{(next 8 terms)} + \text{(next 16 terms)} + \cdots,$$

and then observe that

$$H > 1 + \frac{1}{2} + \left(\frac{1}{4} + \frac{1}{4} \right) + \left(\frac{1}{8} + \frac{1}{8} + \frac{1}{8} + \frac{1}{8} \right) + \cdots,$$

where each new pair of parentheses contains twice as many terms as the previous pair, and each term in a pair is replaced with the last (smallest) term in the pair. The sum of the terms in each modified pair is then $\frac{1}{2}$, and so we have formed a *lower* bound on H that is the sum of an infinite number of $\frac{1}{2}$s. That is, H is "greater than infinity," so to speak, which is just an enthusiastic way of saying H itself blows up.

But that's *not* what happens with S, which can be written as

$$(1.3.12) \quad S = \left(1 + \frac{1}{3^2} + \frac{1}{5^2} + \frac{1}{7^2} + \cdots\right) + \left(\frac{1}{2^2} + \frac{1}{4^2} + \frac{1}{6^2} + \cdots\right).$$

But since

$$\frac{1}{2^2} + \frac{1}{4^2} + \frac{1}{6^2} + \cdots = \frac{1}{(2 \times 1)^2} + \frac{1}{(2 \times 2)^2} + \frac{1}{(2 \times 3)^2} + \cdots$$

$$= \frac{1}{2^2}\left[1 + \frac{1}{2^2} + \frac{1}{3^2} + \frac{1}{4^2} + \frac{1}{5^2} + \cdots\right] = \frac{1}{4}S,$$

then using (1.3.11) and (1.3.12), we have

$$S = \frac{\pi^2}{8} + \frac{1}{4}S.$$

This is easily solved to give the finite sum of

$$(1.3.13) \qquad\qquad S = \left(\frac{4}{3}\right)\left(\frac{\pi^2}{8}\right) = \frac{\pi^2}{6},$$

a result due (via other, more complicated means) to Euler; this discovery (in 1734, 400 years after Oresme) made Euler a superstar in the world of mathematics.[10] Still, *this* derivation is pretty straightforward, depending essentially on nothing much more than the elementary properties of right triangles. With Fourier's approach to the problem, any college freshman today can do in minutes what it took a genius like Euler *years* to do three centuries ago.

Of particular fascination to Euler and his fellow mathematicians must have been that pi is *squared*. We are used to seeing pi, *alone*, in many "ordinary" applications ($2\pi r$ and πr^2, for example, for the circumference and area, respectively, of a circle with radius r), but π^2 was something new. As the English mathematician Augustus de Morgan (1806–1871) is said to have remarked about pi, *alone*, its appearance in mathematics is so common that one imagines "it comes on many occasions through the window and through the door, sometimes even down the chimney." But not pi *squared*.[11]

Here's another quick calculation using Euler's fabulous formula:

$$(e^{it})^n = \{\cos(t) + i\,\sin(t)\}^n = e^{int} = \cos(nt) + i\,\sin(nt).$$

This result,

(1.3.14)
$$\{\cos(t) + i\,\sin(t)\}^n = \cos(nt) + i\,\sin(nt),$$

is called *De Moivre's theorem*,[12] and it is highly useful in both numerical computations and in theoretical analyses. You can find several examples in a previous book of mine[13] of the theorem's value in avoiding lots of grubby numerical work, so let me show you here an application of (1.3.14) in a theoretical context.

In both pure mathematics and physics, the expressions

$$S_1(t) = \sum_{n=1}^{\infty} r^n \cos(nt)$$

and

$$S_2(t) = \sum_{n=1}^{\infty} r^n \sin(nt)$$

often occur, where r is some real number in the interval $0 \le r < 1$. (Do you see *why* this restriction? Think about convergence.) We can find closed-form expressions for these two infinite sums as follows. We start by defining

$$S = \sum_{n=1}^{\infty} z^n = z + z^2 + z^3 + \cdots,$$

with

$$z = r\{\cos(t) + i\,\sin(t)\}.$$

S is simply a geometric series, easily evaluated in the usual way by multiplying through by z. This gives

$$S = \frac{z}{1-z} = \frac{r\cos(t) + i\, r\, \sin(t)}{1 - r\cos(t) - i\, r\, \sin(t)}.$$

Multiplying top and bottom of the right-hand side by the conjugate[14] of the bottom gives

$$S = \frac{\{r\cos(t) + i\, r\, \sin(t)\}\{1 - r\cos(t) + i\, r\, \sin(t)\}}{\{1 - r\cos(t)\}^2 + r^2\sin^2(t)},$$

which reduces to

$$S = \frac{r\cos(t) - r^2 + i\, r\, \sin(t)}{1 - 2r\cos(t) + r^2}.$$

Now, by De Moivre's theorem, we also have

$$S = \sum_{n=1}^{\infty} z^n = \sum_{n=1}^{\infty} r^n\{\cos(t) + i\, \sin(t)\}^n = \sum_{n=1}^{\infty} r^n\{\cos(nt) + i\, \sin(nt)\},$$

and so

$$S = S_1 + iS_2.$$

Equating the imaginary parts of our two results for S gives

$$S_2(t) = \sum_{n=1}^{\infty} r^n\sin(nt) = \frac{r\,\sin(t)}{1 - 2r\cos(t) + r^2}, \quad 0 \le r < 1.$$

And equating real parts, we have

$$S_1(t) = \sum_{n=1}^{\infty} r^n\cos(nt) = \frac{r\cos(t) - r^2}{1 - 2r\cos(t) + r^2}, \quad 0 \le r < 1.$$

With a bit of additional algebra, $S_1(t)$ is often expressed in the alternative form:

$$\frac{r\cos(t)-r^2}{1-2r\cos(t)+r^2} = \frac{1}{2}\left[\frac{2r\cos(t)-2r^2}{1-2r\cos(t)+r^2}\right] = \frac{1}{2}\left[\frac{1-r^2-1+2r\cos(t)-r^2}{1-2r\cos(t)+r^2}\right]$$

$$= \frac{1}{2}\left[\frac{1-r^2}{1-2r\cos(t)+r^2} - \frac{1-2r\cos(t)+r^2}{1-2r\cos(t)+r^2}\right]$$

$$= \frac{1}{2}\left[\frac{1-r^2}{1-2r\cos(t)+r^2} - 1\right],$$

and so we have

$$S_1(t) = \sum_{n=1}^{\infty} r^n\cos(nt) = \frac{1}{2}\left[\frac{1-r^2}{1-2r\cos(t)+r^2}\right] - \frac{1}{2}, \quad 0 \leq r < 1.$$

The quantity

$$\frac{1-r^2}{1-2r\cos(t)+r^2}$$

occurs often enough in advanced mathematics that it has been given its own name: *Poisson's kernel*.[15]

Well, okay, all of this is undeniably fun stuff, but it is relatively lightweight compared to what Fourier did mathematically for physics in *Analytical Theory*. As a quick flip through the rest of this book will show you, there are a *lot* of equations in it. At the beginning of *Analytical Theory*, in what he called a "Preliminary Discourse," Fourier explained to his readers why that was equally so in his book, and his words explain why it is true for this book as well. As he wrote, "Profound study of nature is the most fertile source of mathematical discoveries. . . . Mathematical analysis is as extensive as nature itself." So, all the math you'll read here isn't here because I want to make your life difficult—*it's here because that's the way the world is made.*

To lay the foundation of Fourier's mathematics will take a couple more chapters and so, with no further delay, let's get started.

CHAPTER 2
Fourier's Mathematics

2.1 Fourier Series

As I mentioned in Chapter 1, Fourier's name is today firmly attached to trigonometric series expansions of periodic functions. That is, if T is the period of $f(t)$, which means

$$f(t) = f(t+T),$$

then we'll *assume* (I'll say more about this *assume* business in just a moment) that we can write

$$(2.1.1) \qquad f(t) = \frac{1}{2}a_0 + \sum_{k=1}^{\infty} \left\{ a_k \cos(k\omega_0 t) + b_k \sin(k\omega_0 t) \right\},$$

where

$$\omega_0 = \frac{2\pi}{T}.$$

Although we apply the name of Fourier to such series, they were in fact studied years before he was born, by people like Euler; the Swiss mathematician Daniel Bernoulli (1700–1782), who will appear at a crucial point at the start of Chapter 4; and the French mathematician Jean Le Rond d'Alembert (1717–1783), in connection with physical problems involving the vibrations of strings and the bending of beams under structural loading. For example, in a 1744 letter to a friend, Euler declared that

$$(2.1.2) \qquad \frac{\pi - t}{2} = \sum_{n=1}^{\infty} \frac{\sin(nt)}{n} = \sin(t) + \frac{\sin(2t)}{2} + \frac{\sin(3t)}{3} + \cdots,$$

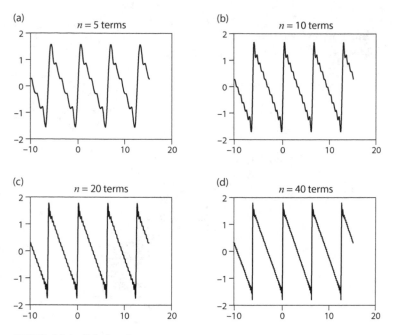

FIGURE 2.1.1. Euler's series.

which was probably the very first "Fourier series," although Euler didn't call it that, since Fourier wouldn't be born until 24 years later. Euler's series gives the same value as $\frac{\pi - t}{2}$ does, for any t in the interval $0 < t < 2\pi$, as shown in Figure 2.1.1. Euler derived this series via a truly outrageously bold use of his "fabulous formula" identity (see Chapter 1), rather than via any *theory* of such series. The following two boxes show how he did it.

If (2.1.1) is to be useful, we obviously have to know what the a_k and b_k coefficients are (the so-called *Fourier coefficients*), and our very first task here will be the derivation of expressions for those coefficients. (When we do that, you'll see why that curious $\frac{1}{2}$ in front of a_0 is there.) So, imagine that we want to express $f(t)$, over the symmetrical interval $-\frac{T}{2} < t < \frac{T}{2}$, as a sum of trigonometric terms with frequencies that are multiples of ω_0.

Euler started with the infinite sum $S(t)$, defined as a geometric series:

$$S(t) = e^{it} + e^{i2t} + e^{i3t} + \cdots .$$

Using the well-known trick for summing such an expression (multiplying through by e^{it}), he wrote

$$e^{it}S(t) = e^{i2t} + e^{i3t} + e^{i4t} + \cdots ,$$

and so

$$S(t) - e^{it}S(t) = e^{it},$$

or

$$S(t) = \frac{e^{it}}{1 - e^{it}} = \frac{e^{it}(1 - e^{-it})}{(1 - e^{it})(1 - e^{-it})} = \frac{e^{it} - 1}{1 - e^{it} - e^{-it} + 1}.$$

Then, using his fabulous formula, it follows that

$$S(t) = \frac{\cos(t) + i\,\sin(t) - 1}{2 - (e^{it} + e^{-it})} = \frac{\cos(t) - 1 + i\,\sin(t)}{2 - 2\cos(t)}$$

$$= \frac{-[1 - \cos(t)] + i\,\sin(t)}{2[1 - \cos(t)]},$$

or

$$S(t) = -\frac{1}{2} + i\left(\frac{1}{2}\right)\frac{\sin(t)}{1 - \cos(t)}.$$

Next, returning to the original expression for $S(t)$ and again using the fabulous formula, Euler wrote

$$S(t) = [\cos(t) + \cos(2t) + \cos(3t) + \cdots]$$

$$+ i[\sin(t) + \sin(2t) + \sin(3t) + \cdots]$$

and then set the real parts of the two expressions for $S(t)$ equal. That is,

$$\cos(t) + \cos(2t) + \cos(3t) + \cdots = -\frac{1}{2}.$$

(Does this make sense if $t = 0$? We'll return to this issue at the end of this section.) In any case, he then indefinitely integrated both sides to get

$$\sin(t) + \frac{\sin(2t)}{2} + \frac{\sin(3t)}{3} + \frac{\sin(4t)}{4} + \cdots = -\frac{1}{2}t + C.$$

He found the value of the arbitrary constant of integration, C, by setting $t = \frac{\pi}{2}$ (*not* $t = 0$, you'll notice) to get

$$1 - \frac{1}{3} + \frac{1}{5} - \frac{1}{7} + \cdots = -\frac{\pi}{4} + C.$$

Recognizing the sum on the left to be equal to $\frac{\pi}{4}$ (a result we derived just after (1.3.7)), the value of C is seen to be $\frac{\pi}{2}$ and so

$$\sin(t) + \frac{\sin(2t)}{2} + \frac{\sin(3t)}{3} + \frac{\sin(4t)}{4} + \cdots = -\frac{1}{2}t + \frac{\pi}{2},$$

or at last, we have Euler's series (2.1.2):

$$\sum_{n=1}^{\infty} \frac{\sin(nt)}{n} = \frac{\pi - t}{2}.$$

The summation on the left is numerically evaluated and plotted in Figure 2.1.1, using the first 5 terms, then the first 10 terms, then the first 20 terms, and finally the first 40 terms. As the number of terms increases, the plots do indeed approach the line $\frac{\pi - t}{2}$ over

(continued)

(continued)

the interval $0 < t < 2\pi$, but the (Gibbs) oscillations around the jump points of discontinuity never disappear. If Euler had actually plotted his series, *he* would have been the discoverer of the Gibbs phenomenon long before Fourier, Wilbraham, and Gibbs were born.

(I'm writing $f(t)$, but of course, everything here works just as well for $f(x)$.) That is, for a *finite* sum of $N+1$ terms, we write

$$S_N(t) = \frac{1}{2}a_0 + \sum_{k=1}^{N}\{a_k\cos(k\omega_0 t) + b_k\sin(k\omega_0 t)\}.$$

If we do this, the obvious question now is: what should the a_k and b_k coefficients be to give the best approximation to $f(t)$?

To answer that question, we have to define what is meant by "best." Here's one way to do that. Define the integral

$$J = \int_{-T/2}^{T/2}[f(t) - S_N(t)]^2\,dt,$$

and then ask what the a_k and b_k coefficients should be to *minimize J*. (*J* is called the *integrated squared error* of the trigonometric approximation.) If we just calculated the integral of the error alone, without squaring it, we could conceivably get a very small *J* even if there are big positive differences between $f(t)$ and $S_N(t)$, over one or more intervals of t, because they are canceled by big *negative* differences between $f(t)$ and $S_N(t)$ over other intervals of t. That is, $S_N(t)$ could be greatly different from $f(t)$ for almost all values of t but still result in a small integrated error. By minimizing the integrated *squared* error, however, such cancelations can't occur. For a *squared* error, a small *J* forces the series approximation to stay "close" to $f(t)$ for almost all values of t.

Now,

$$J = \int_{-T/2}^{T/2}\left[f(t) - \left(\frac{1}{2}a_0 + \sum_{k=1}^{N}\{a_k\cos(k\omega_0 t) + b_k\sin(k\omega_0 t)\}\right)\right]^2\,dt,$$

and we imagine that somehow we have determined the "best" values for all the a_k and all the b_k except for one final a (or one final b). For the sake of a specific demonstration, suppose it is a_n that remains to be determined, where n is in the interval 1 to N. From freshman calculus, then, we wish to determine a_n such that

$$\frac{dJ}{da_n} = 0, \ 1 \le n \le N.$$

We'll treat the $n=0$ case (the value of a_0) separately, at the end of this discussion.

Assuming that the derivative of the integral is the integral of the derivative,[1] we have

$$\frac{dJ}{da_n} = \int_{-T/2}^{T/2} 2 \left[f(t) - \left(\frac{1}{2}a_0 + \sum_{k=1}^{N} \{ a_k \cos(k\omega_0 t) + b_k \sin(k\omega_0 t) \} \right) \right]$$
$$\times \cos(n\omega_0 t) dt,$$

and so, setting this equal to zero, we arrive at

$$\int_{-T/2}^{T/2} f(t) \cos(n\omega_0 t) \, dt$$
$$= \int_{-T/2}^{T/2} \left(\frac{1}{2}a_0 + \sum_{k=1}^{N} \{ a_k \cos(k\omega_0 t) + b_k \sin(k\omega_0 t) \} \right) \cos(n\omega_0 t) \, dt.$$

But since

$$\int_{-T/2}^{T/2} \cos(n\omega_0 t) \, dt = 0,$$

and since

$$\int_{-T/2}^{T/2} \sin(k\omega_0 t) \cos(n\omega_0 t) \, dt = 0,$$

and since

$$\int_{-T/2}^{T/2} \cos(k\omega_0 t) \cos(n\omega_0 t) \, dt = 0 \quad \text{if} \quad k \ne n,$$

and since

$$\int_{-T/2}^{T/2} \cos^2(n\omega_0 t)\,dt = \frac{T}{2},$$

then

$$\int_{-T/2}^{T/2} f(t)\cos(n\omega_0 t)\,dt = a_n \frac{T}{2},$$

or

(2.1.3) $a_n = \dfrac{2}{T}\displaystyle\int_{-T/2}^{T/2} f(t)\cos(n\omega_0 t)\,dt, \ 1 \le n \le N.$

By the same argument,

(2.1.4) $b_n = \dfrac{2}{T}\displaystyle\int_{-T/2}^{T/2} f(t)\sin(n\omega_0 t)\,dt, \ 1 \le n \le N.$

Since we've made no special assumptions about n, then (2.1.3) and (2.1.4) hold for *any* n in the interval $1 \le n \le N$.

Now, what about a_0 (the $n = 0$ coefficient)? We have

$$\frac{dJ}{da_0} = \int_{-T/2}^{T/2} 2\left[f(t) - \left(\frac{1}{2}a_0 + \sum_{k=1}^{N}\{a_k\cos(k\omega_0 t) + b_k\sin(k\omega_0 t)\} \right) \right]$$
$$\times \left(-\frac{1}{2} \right) dt = 0,$$

and so

$$\int_{-T/2}^{T/2} f(t)\,dt = \int_{-T/2}^{T/2} \frac{1}{2}a_0 \,dt = \frac{1}{2}Ta_0,$$

or

(2.1.5) $a_0 = \dfrac{2}{T}\displaystyle\int_{-T/2}^{T/2} f(t)\,dt.$

By including the $\frac{1}{2}$ factor in front of a_0 in (2.1.1), we have arrived at an expression for a_0 that is correctly given by (2.1.3), the expression for a_n, $1 \leq n \leq N$, even when we set $n = 0$ in (2.1.3), and *that* is why mathematicians write $\frac{1}{2}a_0$ instead of just a_0. It's a matter of elegance.

There is a powerful mathematical theorem that says, for the coefficients we have just calculated, not only is J *minimized*, it actually goes to zero as $N \to \infty$ in S_N, *as long as $f(t)$ has just a finite number of discontinuities in a period*. That is a *physical* requirement that is certainly satisfied in any real-world problem. I won't prove that theorem here, but will simply ask that you accept that our mathematician colleagues have, indeed, established its truth.

To end this section on Fourier series, observe that a series will, in general, contain both sine and cosine terms. There are, however, certain special (but highly useful) functions whose Fourier series have only sines (or only cosines). We'll use this fact later in the book, so it's worth taking some time here to discuss how this occurs. First, suppose $f(t)$ is an *even* function over the interval $-\frac{T}{2} < t < \frac{T}{2}$. That is, $f(-t) = f(t)$. Then the integrand in (2.1.3) is even (because $\cos(n\omega_0 t)$ is even, and for functions, even times even is even), while the integrand in (2.1.4) is odd (because $\sin(n\omega_0 t)$ is odd, and for functions, even times odd is odd). Thus, while a_n will in general be nonzero, *all* the b_n coefficients will vanish, and so the Fourier series for an even $f(t)$ will have only cosine terms. In contrast, suppose $f(t)$ is an odd function over the interval $-\frac{T}{2} < t < \frac{T}{2}$. That is, $f(-t) = -f(t)$. Now the opposite situation results, and while in general, the b_n will be nonzero, *all* the a_n coefficients will vanish, and so the Fourier series for an odd $f(t)$ will have only sine terms.[2]

When we evaluate the Fourier series of an $f(t)$ defined on the interval $-\frac{T}{2} < t < \frac{T}{2}$, the result will *not* equal $f(\hat{t})$ for a $\hat{t}$ not in the interval $-\frac{T}{2}$ to $\frac{T}{2}$. The series will indeed *converge* for that $\hat{t}$, but not to $f(\hat{t})$ but rather to the value of what is called the *periodic extension* of $f(t)$ up and down the t-axis. Figure 2.1.2 shows the periodic extension (for $T = 2\pi$) of an even function (t^2), and Figure 2.1.3 shows the periodic extension (for $T = 2\pi$) of an odd function (t).

As an example of all this, let's find the Fourier series of $f(t) = t^2$, $-\pi \leq t \leq \pi$. First, notice that since $f(t)$ is an even function, there will be

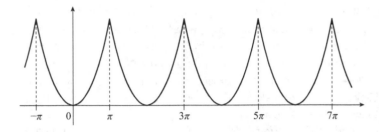

FIGURE 2.1.2. The periodic extension of an even function.

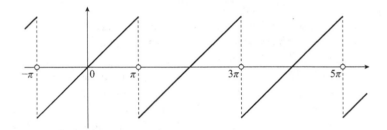

FIGURE 2.1.3. The periodic extension of an odd function.

only cosine terms in the series. That is, all the Fourier b-coefficients will be zero, and we need calculate only the a-coefficients. So, as $T = 2\pi$ (and thus $\omega_0 = 1$), we have

$$a_n = \frac{1}{\pi} \int_{-\pi}^{\pi} t^2 \cos(nt)\,dt = \frac{2}{\pi} \int_0^{\pi} t^2 \cos(nt)\,dt,$$

an integral easily done by parts (even more easily by looking in a math table!). In any case, for $n \geq 1$, we have

$$a_n = \frac{2}{\pi} \left[\frac{2t}{n^2} \cos(nt) + \left(\frac{t^2}{n} - \frac{2}{n^3} \right) \sin(nt) \right] \Bigg|_0^{\pi}$$

$$= \frac{4}{n^2 \pi} [\pi \cos(n\pi)] = \frac{4}{n^2} \cos(n\pi) = \frac{4}{n^2} (-1)^n,$$

while for $n = 0$,

$$a_0 = \frac{1}{\pi} \int_{-\pi}^{\pi} t^2 \, dt = \frac{2}{\pi} \int_0^{\pi} t^2 \, dt = \frac{2}{\pi} \left(\frac{1}{3} t^3 \right) \Big|_0^{\pi} = \frac{2\pi^2}{3}.$$

Thus, we obtain

$$(2.1.6) \qquad t^2 = \frac{\pi^2}{3} + 4 \sum_{n=1}^{\infty} (-1)^n \frac{\cos(nt)}{n^2},$$

$$= \frac{\pi^2}{3} - 4 \left[\cos(t) - \frac{\cos(2t)}{2^2} + \frac{\cos(3t)}{3^2} - \cdots \right]$$

which is valid for $-\pi \le t \le \pi$.

In particular, if we set $t = \pi$ in (2.1.6), then our series reduces to

$$\pi^2 = \frac{\pi^2}{3} - 4 \left[\cos(\pi) - \frac{\cos(2\pi)}{2^2} + \frac{\cos(3\pi)}{3^2} - \cdots \right],$$

or

$$\frac{2}{3} \pi^2 = 4 \left[1 + \frac{1}{2^2} + \frac{1}{3^2} + \cdots \right],$$

and so

$$1 + \frac{1}{2^2} + \frac{1}{3^2} + \cdots = \frac{2}{12} \pi^2 = \frac{1}{6} \pi^2,$$

which is (1.3.12), derived in Chapter 1 with nothing but AP high school math, and by Euler using other (*far* more complicated) means. It is reassuring to see that Fourier's math is consistent with a result obtained using entirely different methods. See the next box for another interesting implication of (2.1.6).

Now, just to be absolutely sure you are suitably impressed with the power of Fourier series, let me show you two more astonishing calculations. To set you up for the first one, we need to do some additional (but still pretty straightforward) analysis. Repeating (2.1.1),

$$f(t) = \frac{1}{2}a_0 + \sum_{k=1}^{\infty}\{a_k \cos(k\omega_0 t) + b_k \sin(k\omega_0 t)\},$$

where $\omega_0 T = 2\pi$. Using Euler's identity, this becomes

$$f(t) = \frac{1}{2}a_0 + \sum_{k=1}^{\infty}\left\{ a_k \frac{e^{ik\omega_0 t} + e^{-ik\omega_0 t}}{2} + b_k \frac{e^{ik\omega_0 t} - e^{-ik\omega_0 t}}{2i} \right\}$$

$$= \frac{1}{2}a_0 + \sum_{k=1}^{\infty}\left[\left\{ \frac{a_k}{2} + \frac{b_k}{2i} \right\} e^{ik\omega_0 t} + \left\{ \frac{a_k}{2} - \frac{b_k}{2i} \right\} e^{-ik\omega_0 t} \right].$$

If we let the summation index run from minus to plus infinity, then we can write a *complex* Fourier series as

$$(2.1.7) \qquad f(t) = \sum_{k=-\infty}^{\infty} c_k e^{ik\omega_0 t}, \quad \omega_0 T = 2\pi,$$

where the c_k are constants (in general, *complex-valued* constants).

Rearranging (2.1.6) just a bit gives

$$\cos(t) - \frac{\cos(2t)}{2^2} + \frac{\cos(3t)}{3^2} - \cdots = \frac{\frac{\pi^2}{3} - t^2}{4} = \frac{\pi^2 - 3t^2}{12},$$

or

$$\frac{\pi^2}{12} - \frac{t^2}{4} = \sum_{n=1}^{\infty} (-1)^{n-1} \frac{\cos(nt)}{n^2}.$$

Integrating indefinitely,

$$\frac{\pi^2}{12}t - \frac{t^3}{12} + C = \sum_{n=1}^{\infty} (-1)^{n-1} \frac{\sin(nt)}{n^3},$$

or setting $t=0$ to evaluate C, we have $0+C=0$, and so

$$\sum_{n=1}^{\infty}(-1)^{n-1}\frac{\sin(nt)}{n^3}=\frac{\pi^2 t-t^3}{12}, \quad -\pi \leq t \leq \pi.$$

Setting $t=\dfrac{\pi}{2}$, this becomes the statement

$$1-\frac{1}{3^3}+\frac{1}{5^3}-\frac{1}{7^3}+\cdots=\frac{\pi^3}{32},$$

which should remind you of note 10 in Chapter 1.

Let's next suppose that $f(t)$ is a real-valued function. Since the conjugate[3] of a real value is the real value, then

$$f(t)=\sum_{k=-\infty}^{\infty}c_k e^{ik\omega_0 t}=f^*(t)=\left\{\sum_{k=-\infty}^{\infty}c_k e^{ik\omega_0 t}\right\}^*.$$

Since the conjugate of a sum is the sum of the conjugates, and since the conjugate of a product is the product of the conjugates,[4] then

(2.1.8) $$\sum_{k=-\infty}^{\infty}c_k e^{ik\omega_0 t}=\sum_{k=-\infty}^{\infty}c_k^* e^{-ik\omega_0 t},$$

which tells us that if $f(t)$ is a real-valued function, then $c_{-k}=c_k^*$ (to see this, set the coefficients of matching exponential terms on each side of (2.1.8) equal to each other). Notice, too, that for the case of $k=0$ we have $c_0=c_0^*$, which says that, for any real-valued function $f(t)$, we'll have c_0 always come out as real-valued.

Now, for some (any) particular integer from minus to plus infinity (let's say, n), multiply both sides of (2.1.7) by $e^{-in\omega_0 t}$ and integrate over a period, that is, over any interval of length T. Then, with t' an arbitrary (but fixed) value of t,

$$\int_{t'}^{t'+T}f(t)e^{-in\omega_0 t}\,dt=\int_{t'}^{t'+T}\left\{\sum_{k=-\infty}^{\infty}c_k e^{ik\omega_0 t}\right\}e^{-in\omega_0 t}\,dt,$$

or

(2.1.9) $$\int_{t'}^{t'+T} f(t)e^{-in\omega_0 t}\, dt = \sum_{k=-\infty}^{\infty} c_k \int_{t'}^{t'+T} e^{i(k-n)\omega_0 t}\, dt.$$

The integral on the right in (2.1.9) is easy to do, and we'll do it in two steps. Once for $k \neq n$, and then again for $k = n$. So, if $k \neq n$, we have

$$\int_{t'}^{t'+T} e^{i(k-n)\omega_0 t}\, dt = \left\{ \frac{e^{i(k-n)\omega_0 t}}{i(k-n)\omega_0} \right\}\Bigg|_{t'}^{t'+T} = \frac{e^{i(k-n)\omega_0(t'+T)} - e^{i(k-n)\omega_0 t'}}{i(k-n)\omega_0}$$

$$= \frac{e^{i(k-n)\omega_0 t'}\{e^{i(k-n)\omega_0 T} - 1\}}{i(k-n)\omega_0}.$$

Since $\omega_0 T = 2\pi$, since $k - n$ is a nonzero integer, and since Euler's identity tells us that $e^{i(k-n)\omega_0 T} = 1$, then the integral is zero for the $k \neq n$ case. For the $k = n$ case, the integral becomes

$$\int_{t'}^{t'+T} e^0\, dt = \{t\}\Big|_{t'}^{t'+T} = T.$$

So, in summary,

(2.1.10) $$\int_{period} e^{i(k-n)\omega_0 t}\, dt = \begin{cases} 0, & k \neq n \\ T, & k = n \end{cases}.$$

Thus, (2.1.9) becomes

$$\int_{period} f(t)e^{-in\omega_0 t}\, dt = c_n T.$$

So for all k, the Fourier coefficients in (2.1.7) are given by

(2.1.11) $$c_k = \frac{1}{T}\int_{period} f(t)e^{-ik\omega_0 t}\, dt, \quad \omega_0 T = 2\pi.$$

(just replace n with k on both sides of the last integral). Okay, that's the end of the "preliminary analysis" I mentioned. Now, let's do something amazing with it.

The *energy* (a term used by mathematicians, physicists, and engineers alike) of the real-valued function $f(t)$, over a period, is defined to be the integral

$$W = \int_{period} f^2(t)\, dt.$$

(The reason for calling W the *energy* of $f(t)$ is that if $f(t)$ is the voltage drop across a 1-ohm resistor, then W is the electrical energy dissipated—as heat—by the resistor during one period. If this is not clear right now, don't worry—we'll return to this point in Chapter 5, where it *will* become clear.) If we substitute the complex Fourier series for $f(t)$ into the energy integral, writing $f^2(t) = f(t)f(t)$ and using a different index of summation for each $f(t)$, we get

$$W = \int_{period} \left\{ \sum_{m=-\infty}^{\infty} c_m e^{im\omega_0 t} \right\} \left\{ \sum_{n=-\infty}^{\infty} c_n e^{in\omega_0 t} \right\} dt,$$

or

(2.1.12)
$$W = \sum_{m=-\infty}^{\infty} \sum_{n=-\infty}^{\infty} c_m c_n \int_{period} e^{i(m+n)\omega_0 t}\, dt.$$

This last integral is one we've already done, back in (2.1.10). That is, the integral is zero when $m + n \neq 0$, and it is T when $m + n = 0$ (when $m = -n$). So, remembering that for a real-valued $f(t)$, $c_{-k} = c_k^*$, (2.1.12) reduces to

$$W = \sum_{k=-\infty}^{\infty} c_k c_{-k} T = T \sum_{k=-\infty}^{\infty} c_k c_k^* = T \sum_{k=-\infty}^{\infty} \left| c_k \right|^2,$$

or remembering that c_0 is real since $f(t)$ is real,

(2.1.13)
$$\frac{W}{T} = \frac{1}{T} \int_{period} f^2(t)\, dt = c_0^2 + 2 \sum_{k=1}^{\infty} \left| c_k \right|^2,$$

a result called *Parseval's power formula*.[5] (Energy per unit time—the units of $\frac{W}{T}$—is *power* in physics lingo.)

As a dramatic illustration of the use of (2.1.13), suppose that

$$f(t) = e^{-pt}, \quad 0 < t < T = 2\pi, \quad \text{with } p \text{ an arbitrary positive constant,}$$

which defines a single period of a function extended over the entire t-axis. Then, from (2.1.11), and observing that $\omega_0 = 1$, we find

$$c_k = \frac{1}{2\pi} \int_0^{2\pi} e^{-pt} e^{-ikt}\, dt = \frac{1}{2\pi} \int_0^{2\pi} e^{-(p+ik)t}\, dt = \frac{1}{2\pi} \left\{ \frac{e^{-(p+ik)t}}{-(p+ik)} \right\} \Bigg|_0^{2\pi}$$

$$= \left(\frac{1}{2\pi} \right) \frac{1 - e^{-(p+ik)2\pi}}{p+ik} = \left(\frac{1}{2\pi} \right) \frac{1 - e^{-2\pi p} e^{-ik2\pi}}{p+ik} = \frac{1 - e^{-2\pi p}}{2\pi(p+ik)},$$

because $e^{-ik2\pi} = 1$ for all integers k. Thus,

$$|c_k|^2 = \frac{(1 - e^{-2\pi p})^2}{4\pi^2(p^2 + k^2)}, \quad c_0 = \frac{1 - e^{-2\pi p}}{2\pi p}.$$

Therefore, (2.1.13) says

$$\frac{1}{2\pi} \int_0^{2\pi} e^{-2pt}\, dt = \left(\frac{1 - e^{-2\pi p}}{2\pi p} \right)^2 + 2 \sum_{k=1}^{\infty} \frac{(1 - e^{-2\pi p})^2}{4\pi^2(p^2 + k^2)},$$

or

$$(2.1.14) \quad \frac{1}{2\pi(1 - e^{-2\pi p})^2} \int_0^{2\pi} e^{-2pt}\, dt = \frac{1}{4\pi^2 p} + \frac{1}{2\pi^2} \sum_{k=1}^{\infty} \frac{1}{p^2 + k^2}.$$

The integral on the left in (2.1.14) is easy to do:

$$\int_0^{2\pi} e^{-2pt}\, dt = \left(\frac{e^{-2pt}}{-2p} \right) \Bigg|_0^{2\pi} = \frac{1 - e^{-4\pi p}}{2p} = \frac{(1 - e^{-2\pi p})(1 + e^{-2\pi p})}{2p},$$

and so (2.1.14) becomes

$$\frac{1 + e^{-2\pi p}}{4\pi p(1 - e^{-2\pi p})} = \frac{1}{4\pi^2 p^2} + \frac{1}{2\pi^2} \sum_{k=1}^{\infty} \frac{1}{p^2 + k^2},$$

or

$$\sum_{k=1}^{\infty} \frac{1}{p^2+k^2} = 2\pi^2 \left[\frac{1+e^{-2\pi p}}{4\pi p(1-e^{-2\pi p})} - \frac{1}{4\pi^2 p^2} \right],$$

or at last,

(2.1.15)
$$\sum_{k=1}^{\infty} \frac{1}{p^2+k^2} = \left(\frac{\pi}{2p} \right) \frac{1+e^{-2\pi p}}{1-e^{-2\pi p}} - \frac{1}{2p^2}.$$

Pretty, yes, but is (2.1.15) *correct*? One quick, *easy* way to tremendously enhance our confidence in (2.1.15) is to simply evaluate, *numerically*, both sides for some various values of p and then just look to see whether we always get the same answers from each side. Pure mathematicians generally claim to be not very impressed with this approach (in public, anyway, although I suspect they privately like the numbers to agree just as much as everybody else does): engineers and physicists are more easily persuaded by such calculations. Anyway, here are some numerical results, where the sum on the left in (2.1.15) was calculated using the first 10,000 terms of the summation:

p	left side of (2.1.15)	right side of (2.1.15)
0.1	1.63411 . . .	1.63421 . . .
1	1.07657 . . .	1.07667 . . .
2	0.66030 . . .	0.66040 . . .
5	0.29405 . . .	0.29415 . . .

The numerical agreement looks pretty good, in my opinion.

You'll notice, too, that as p decreases toward zero, both sides of (2.1.15) seem to be approaching the value of $\frac{\pi^2}{6} = 1.644934\ldots$, the value of the sum for $p = 0$ (the sum, of course, is then Euler's famous sum of the reciprocals of the positive integers squared found in (1.3.12)). This suggests another, now theoretical, test for (2.1.15): is it true that

$$\lim_{p\to 0} \left\{ \left(\frac{\pi}{2p} \right) \frac{1+e^{-2\pi p}}{1-e^{-2\pi p}} - \frac{1}{2p^2} \right\} = \frac{\pi^2}{6}?$$

The answer is *yes,* and I'll leave the demonstration of that as a challenge calculation for you.[6]

Now, to wrap this section up with the second astonishing Fourier series calculation I promised you earlier, I'll next show you an *un-bounded* periodic function that can be written as a Fourier series. As a prelude to that calculation, I'll start by first deriving a result we'll eventually need to complete the Fourier calculations. Consider the *finite* sum

$$S = \sum_{n=1}^{N} \cos(n\theta) = \cos(\theta) + \cos(2\theta) + \cdots + \cos(N\theta), N = 1, 2, 3, \cdots.$$

Using Euler's identity, we can write S as

$$
\begin{aligned}
S &= \frac{e^{i\theta} + e^{-i\theta}}{2} + \frac{e^{i2\theta} + e^{-i2\theta}}{2} + \cdots + \frac{e^{iN\theta} + e^{-iN\theta}}{2} \\
&= \frac{1}{2}(e^{i\theta} + e^{i2\theta} + \cdots + e^{iN\theta}) + \frac{1}{2}(e^{-i\theta} + e^{-i2\theta} + \cdots + e^{-iN\theta}) \\
&= \frac{1}{2}S_1 + \frac{1}{2}S_2,
\end{aligned}
$$

where

$$S_1 = e^{i\theta} + e^{i2\theta} + \cdots + e^{iN\theta}, \quad S_2 = e^{-i\theta} + e^{-i2\theta} + \cdots + e^{-iN\theta}.$$

Therefore,

$$S_1 - e^{i\theta} S_1 = e^{i\theta} - e^{i(N+1)\theta},$$

or

$$S_1 = \frac{e^{i\theta} - e^{i(N+1)\theta}}{1 - e^{i\theta}}.$$

In the same way, we find

$$S_2 = \frac{e^{-i\theta} - e^{-i(N+1)\theta}}{1 - e^{-i\theta}}.$$

So

$$\sum_{n=1}^{N} \cos(n\theta) = \frac{1}{2}\left[\frac{e^{i\theta} - e^{i(N+1)\theta}}{1 - e^{i\theta}} + \frac{e^{-i\theta} - e^{-i(N+1)\theta}}{1 - e^{-i\theta}}\right]$$

$$= \frac{1}{2}\left[\frac{e^{i\frac{1}{2}\theta}\left\{e^{i\frac{1}{2}\theta} - e^{i\left(N+\frac{1}{2}\right)\theta}\right\}}{e^{i\frac{1}{2}\theta}\left\{e^{-i\frac{1}{2}\theta} - e^{i\frac{1}{2}\theta}\right\}} + \frac{e^{-i\frac{1}{2}\theta}\left\{e^{-i\frac{1}{2}\theta} - e^{-i\left(N+\frac{1}{2}\right)\theta}\right\}}{e^{-i\frac{1}{2}\theta}\left\{e^{i\frac{1}{2}\theta} - e^{-i\frac{1}{2}\theta}\right\}}\right]$$

$$= \frac{1}{2}\left[\frac{e^{i\frac{1}{2}\theta} - e^{i\left(N+\frac{1}{2}\right)\theta}}{-i2\sin\left\{\frac{1}{2}\theta\right\}} + \frac{e^{-i\frac{1}{2}\theta} - e^{-i\left(N+\frac{1}{2}\right)\theta}}{i2\sin\left\{\frac{1}{2}\theta\right\}}\right]$$

$$= \frac{1}{2}\left[\frac{-e^{i\frac{1}{2}\theta} + e^{i\left(N+\frac{1}{2}\right)\theta} + e^{-i\frac{1}{2}\theta} - e^{-i\left(N+\frac{1}{2}\right)\theta}}{i2\sin\left\{\frac{1}{2}\theta\right\}}\right]$$

$$= \frac{1}{2}\left[\frac{-i2\sin\left\{\frac{1}{2}\theta\right\}}{i2\sin\left\{\frac{1}{2}\theta\right\}} + \frac{i2\sin\left\{\left(N+\frac{1}{2}\right)\theta\right\}}{i2\sin\left\{\frac{1}{2}\theta\right\}}\right],$$

and therefore,

$$(2.1.16) \qquad \sum_{n=1}^{N} \cos(n\theta) = -\frac{1}{2} + \frac{\sin\left\{\left(N+\frac{1}{2}\right)\theta\right\}}{2\sin\left\{\frac{1}{2}\theta\right\}},$$

a result that is sometimes referred to as *Lagrange's identity* (after Fourier's colleague in the French Academy of Sciences). As a partial check on our work, notice that if $\theta = 0$, the left-hand side of (2.1.16) reduces to $\sum_{n=1}^{N} 1 = N$, while the right-hand side (using L'Hôpital's rule[7] for the resulting indeterminate ratio $\frac{0}{0}$) is

$$-\frac{1}{2}+\frac{N+\frac{1}{2}}{2\left(\frac{1}{2}\right)}=-\frac{1}{2}+N+\frac{1}{2}=N,$$

in agreement. (At this point, look back to the first shaded box of this chapter, at Euler's wild derivation of the series for $\frac{\pi-t}{2}$.)

If we integrate both sides of (2.1.16) from $-\pi$ to π, and notice that all the resulting cosine integrals on the left vanish, we have

$$0=-\int_{-\pi}^{\pi}\frac{1}{2}d\theta+\int_{-\pi}^{\pi}\frac{\sin\left\{\left(N+\frac{1}{2}\right)\theta\right\}}{2\sin\left\{\frac{1}{2}\theta\right\}}d\theta,$$

and because the integrand of the last integral is even, we get

$$\pi=2\int_{0}^{\pi}\frac{\sin\left\{\left(N+\frac{1}{2}\right)\theta\right\}}{2\sin\left\{\frac{1}{2}\theta\right\}}d\theta,$$

or

(2.1.17)
$$\frac{1}{\pi}\int_{0}^{\pi}\frac{\sin\left\{\left(N+\frac{1}{2}\right)\theta\right\}}{2\sin\left\{\frac{1}{2}\theta\right\}}d\theta=\frac{1}{2}.$$

In our initial definition of S, N can be any *positive* integer but, in fact, (2.1.17) obviously holds for the case of $N=0$ as well. Because of this, it immediately follows that

(2.1.18)
$$\frac{1}{\pi}\int_{0}^{\pi}\frac{\sin\left\{\left(N-\frac{1}{2}\right)\theta\right\}}{2\sin\left\{\frac{1}{2}\theta\right\}}d\theta=\frac{1}{2}$$

for N any *positive* integer. (Note *carefully* that (2.1.18) does *not* hold for the $N = 0$ case.) We'll come back to (2.1.17) and (2.1.18) in just a bit.

Now we are ready for the unbounded function of (2.1.19), shown in Figure 2.1.4:

$$(2.1.19) \qquad f(t) = -\ln\left|2\sin\left(\frac{t}{2}\right)\right|, \quad -\infty < t < \infty.$$

This function blows up at all values of t equal to any integer multiple of 2π (including $t = 0$). (Don't overlook the absolute value signs that keep the argument of the log function from being negative, which would make the log function imaginary.) The period of $f(t)$ is clearly 2π. Since $f(t)$ is an even function, we immediately know that the b_n Fourier coefficients of (2.1.4) vanish. And since $T = 2\pi$ (and so $\omega_0 = 1$), (2.1.3) tells us that

$$a_n = -\frac{2}{2\pi}\int_{-\pi}^{\pi}\ln\left|2\sin\left(\frac{t}{2}\right)\right|\cos(nt)\,dt$$

$$= -\frac{1}{\pi}\int_{-\pi}^{\pi}\ln\left|2\sin\left(\frac{t}{2}\right)\right|\cos(nt)\,dt, \quad n = 0, 1, 2, \ldots.$$

Because $\cos(nt)$, like $f(t)$, is even, the integrand of this last integral is even, and so we can write the integral as twice that when done over just half the original integration interval:

$$(2.1.20) \qquad a_n = -\frac{2}{\pi}\int_0^{\pi}\ln\left\{2\sin\left(\frac{t}{2}\right)\right\}\cos(nt)\,dt, \quad n = 0, 1, 2, \ldots,$$

where the absolute value signs have been dropped, since the argument of the log function is never negative over the revised interval of integration.

For the $n = 0$ case (that is, for a_0), we have

$$(2.1.21) \qquad a_0 = -\frac{2}{\pi}\int_0^{\pi}\ln\left\{2\sin\left(\frac{t}{2}\right)\right\}dt = -\frac{2}{\pi}I.$$

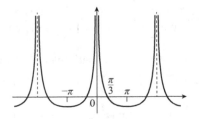

FIGURE 2.1.4. A periodic, unbounded function.

The integral I certainly looks, at first glance, like it might be a tough nut to crack, but in fact, it yields to the following direct calculation:

$$I = \int_0^\pi \ln\left\{2\sin\left(\frac{t}{2}\right)\right\} dt = \int_0^\pi \left[\ln(2) + \ln\left\{\sin\left(\frac{t}{2}\right)\right\}\right] dt,$$

and so

$$(2.1.22) \qquad I = \pi \ln(2) + \int_0^\pi \ln\left\{\sin\left(\frac{t}{2}\right)\right\} dt = \pi \ln(2) + J,$$

where

$$(2.1.23) \qquad J = \int_0^\pi \ln\left\{\sin\left(\frac{t}{2}\right)\right\} dt.$$

We evaluate J by making the change of variable $u = \frac{t}{2}$ (and so $dt = 2du$). Then (2.1.23) becomes

$$J = 2\int_0^{\pi/2} \ln\{\sin(u)\} du,$$

or remembering the identity $\sin(u) = 2\sin\left(\frac{u}{2}\right)\cos\left(\frac{u}{2}\right)$, we have

$$J = 2\int_0^{\pi/2} \ln\left\{2\sin\left(\frac{u}{2}\right)\cos\left(\frac{u}{2}\right)\right\} du$$

$$= 2\int_0^{\pi/2}\left[\ln(2) + \ln\left\{\sin\left(\frac{u}{2}\right)\right\} + \ln\left\{\cos\left(\frac{u}{2}\right)\right\}\right] du,$$

which says

$$J = \pi\ln(2) + 2\int_0^{\pi/2}\ln\left\{\sin\left(\frac{u}{2}\right)\right\} du + 2\int_0^{\pi/2}\ln\left\{\cos\left(\frac{u}{2}\right)\right\} du.$$

In the last integral, change the variable to $u = \pi - s$ $(du = -ds)$, and so

$$2\int_0^{\pi/2}\ln\left\{\cos\left(\frac{u}{2}\right)\right\} du = 2\int_\pi^{\pi/2}\ln\left\{\cos\left(\frac{\pi-s}{2}\right)\right\}(-ds)$$

$$= 2\int_{\pi/2}^\pi\ln\left\{\sin\left(\frac{s}{2}\right)\right\} ds = 2\int_{\pi/2}^\pi\ln\left\{\sin\left(\frac{u}{2}\right)\right\} du,$$

where I've used the identity $\cos\left(\dfrac{\pi-s}{2}\right) = \sin\left(\dfrac{s}{2}\right)$. Thus, we have

$$J = \pi\ln(2) + 2\int_0^{\pi/2}\ln\left\{\sin\left(\frac{u}{2}\right)\right\} du + 2\int_{\pi/2}^\pi\ln\left\{\sin\left(\frac{u}{2}\right)\right\} du,$$

or

$$J = \pi\ln(2) + 2\int_0^\pi\ln\left\{\sin\left(\frac{u}{2}\right)\right\} du$$

and so, recalling (2.1.23),

$$J = \pi\ln(2) + 2J.$$

This says $J = -\pi\ln(2)$, and putting that into (2.1.22), we get $I = 0$. Thus, after all our work, (2.1.21) tells us that

$$a_0 = 0,$$

which is, I think, a bit of a surprise (it certainly isn't obvious—not to me, anyway—from Figure 2.1.4 that the average value of $f(t)$, over a period, is zero. (A look at (2.1.5) shows that $\frac{1}{2}a_0$ *is the average over a period of $f(t)$.*)

For the a_n coefficients when $n \geq 1$, return to (2.1.20) and integrate by parts. That is, to evaluate the integral

$$\int_0^\pi \ln\left\{2\sin\left(\frac{t}{2}\right)\right\}\cos(nt)dt,$$

we'll use the integration-by-parts formula

$$\int_0^\pi u\ dv = (uv)\Big|_0^\pi - \int_0^\pi v\ du$$

with

$$u = \ln\left\{2\sin\left(\frac{t}{2}\right)\right\},\ dv = \cos(nt)dt.$$

This gives us

$$du = \frac{2\cos\left(\dfrac{t}{2}\right)\dfrac{1}{2}}{2\sin\left(\dfrac{t}{2}\right)}dt = \frac{\cos\left(\dfrac{t}{2}\right)}{2\sin\left(\dfrac{t}{2}\right)}dt,\quad v = \frac{1}{n}\sin(nt).$$

Thus, for $n > 0$, we have

$$\int_0^\pi \ln\left\{2\sin\left(\frac{t}{2}\right)\right\}\cos(nt)dt$$

$$= \left\{\frac{\ln\left\{2\sin\left(\dfrac{t}{2}\right)\right\}\sin(nt)}{n}\right\}\Bigg|_0^\pi - \frac{1}{2n}\int_0^\pi \frac{\sin(nt)\cos\left(\dfrac{t}{2}\right)}{\sin\left(\dfrac{t}{2}\right)}dt,$$

or, because the term on the right in the curly brackets vanishes at both limits,[8]

$$\int_0^\pi \ln\left\{2\sin\left(\frac{t}{2}\right)\right\}\cos(nt)\,dt = -\frac{1}{2n}\int_0^\pi \frac{\sin(nt)\cos\left(\dfrac{t}{2}\right)}{\sin\left(\dfrac{t}{2}\right)}\,dt.$$

Thus,

$$a_n = \frac{1}{\pi n}\int_0^\pi \frac{\sin(nt)\cos\left(\dfrac{t}{2}\right)}{\sin\left(\dfrac{t}{2}\right)}\,dt, \quad n = 1, 2, 3, \ldots.$$

Now, recall the identity

$$\sin(nt)\cos\left(\frac{t}{2}\right) = \frac{1}{2}\left[\sin\left\{\left(n+\frac{1}{2}\right)t\right\} + \sin\left\{\left(n-\frac{1}{2}\right)t\right\}\right].$$

This says

$$(2.1.24) \quad a_n = \frac{1}{n}\left[\frac{1}{\pi}\int_0^\pi \frac{\sin\left\{\left(n+\dfrac{1}{2}\right)t\right\}}{2\sin\left(\dfrac{t}{2}\right)}\,dt + \frac{1}{\pi}\int_0^\pi \frac{\sin\left\{\left(n-\dfrac{1}{2}\right)t\right\}}{2\sin\left(\dfrac{t}{2}\right)}\,dt\right],$$

and if you look back at our preliminary results in (2.1.17) and (2.1.18), you'll see that each of these integrals (including the $\frac{1}{\pi}$ factors), is equal to $\frac{1}{2}$. Thus we see (2.1.24) reduce to

$$(2.1.25) \qquad\qquad a_n = \frac{1}{n}, \quad n > 0,$$

and so the Fourier series for (2.1.19) is

$$(2.1.26) \quad -\ln\left|2\sin\left(\frac{t}{2}\right)\right| = \sum_{n=1}^{\infty} \frac{\cos(nt)}{n}$$

$$= \cos(t) + \frac{\cos(2t)}{2} + \frac{\cos(3t)}{3} + \cdots.$$

Figure 2.1.5 shows (2.1.26) using the first 5, first 10, first 25, and then the first 50 terms.

Our analysis may remind you of our earlier discussion of Euler's wild derivation of (2.1.2), in the first two shaded boxes of this chapter, of the series for $\frac{\pi - t}{2}$, in which all the cosines of (2.1.26) are replaced with sines. But of course, a big difference results from simply shifting each of Euler's sines by $\frac{\pi}{2}$ radians into a cosine. First of all, the right-hand side of (2.1.26) does indeed blow up if t is any integer multiple of 2π, as the trigonometric series becomes the harmonic series. The blow-up is a *very* slow one, of course, as the harmonic series diverges only as the logarithm of the number of terms.

It is also interesting to note that setting $t = \pi$ in (2.1.26) gives us another famous result:

$$-\ln\left|2\sin\left(\frac{\pi}{2}\right)\right| = -\ln(2) = \cos(\pi) + \frac{\cos(2\pi)}{2} + \frac{\cos(3\pi)}{3} + \frac{\cos(4\pi)}{4} + \cdots$$

$$= -1 + \frac{1}{2} - \frac{1}{3} + \frac{1}{4} - \cdots,$$

or

$$\ln(2) = 1 - \frac{1}{2} + \frac{1}{3} - \frac{1}{4} + \cdots,$$

a result due to Mercator that I mentioned back in Chapter 1 (see just before (1.3.8)). A more direct calculation of Mercator's series, which you may have already noticed, is simply to write

$$\int \frac{dt}{1+t} = \ln(1+t) + C = \int (1 - t + t^2 - t^3 + \cdots)dt$$

$$= t - \frac{1}{2}t^2 + \frac{1}{3}t^3 - \frac{1}{4}t^4 + \cdots$$

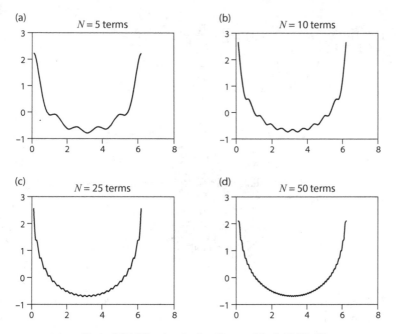

FIGURE 2.1.5. Plots of (2.1.26) using the first N terms ($N=5$, 10, 25, 50).

and then observe that, at $t=0$, $\ln(1)+C=0$, or $C=-\ln(1)=0$. Thus

$$\ln(1+t) = t - \frac{1}{2}t^2 + \frac{1}{3} - \frac{1}{4}t^4 + \cdots,$$

and so, setting $t=1$, we have Mercator's series. This is much shorter than was our derivation of (2.1.26), but of course, (2.1.26) is a more general and far deeper result.

2.2 Fourier Transforms

At this point, I hope I've convinced you of the power of Fourier series. Now we're ready to push the series even further into new territory, and to that end, let's return to (2.1.3) and (2.1.4) and write $T=2l$, so the Fourier coefficients become

(2.2.1) $$a_n = \frac{1}{l}\int_{-l}^{l} f(t)\cos\left(\frac{n\pi t}{l}\right)dt, \quad n=0,1,2,\ldots,$$

and

(2.2.2) $$b_n = \frac{1}{l}\int_{-l}^{l} f(t)\sin\left(\frac{n\pi t}{l}\right)dt, \quad n=1,2,\ldots,$$

because $\omega_0 T = 2\pi$ says that, with $T = 2l$, we have $\omega_0 = \frac{\pi}{l}$. The Fourier series for $f(t)$ is then

$$f(t) = \frac{1}{2}a_0 + \sum_{n=1}^{\infty}\left\{a_n\cos\left(\frac{n\pi t}{l}\right) + b_n\sin\left(\frac{n\pi t}{l}\right)\right\}.$$

Putting in what a_n and b_n are, we have (with u as a dummy variable of integration)

$$f(t) = \frac{1}{2l}\int_{-l}^{l} f(u)\,du + \sum_{n=1}^{\infty}\cos\left(\frac{n\pi t}{l}\right)\frac{1}{l}\int_{-l}^{l} f(u)\cos\left(\frac{n\pi u}{l}\right)du$$

$$+ \sum_{n=1}^{\infty}\sin\left(\frac{n\pi t}{l}\right)\frac{1}{l}\int_{-l}^{l} f(u)\sin\left(\frac{n\pi u}{l}\right)du,$$

or

(2.2.3) $$f(t) = \frac{1}{2l}\int_{-l}^{l} f(u)du$$

$$+ \sum_{n=1}^{\infty}\frac{1}{l}\int_{-l}^{l} f(u)\left\{\cos\left(\frac{n\pi t}{l}\right)\cos\left(\frac{n\pi u}{l}\right)\right.$$

$$\left. + \sin\left(\frac{n\pi t}{l}\right)\sin\left(\frac{n\pi u}{l}\right)\right\}du.$$

Next, recall the trigonometric identity

$$\cos(A)\cos(B) + \sin(A)\sin(B) = \cos(A - B),$$

and so, with $A = \dfrac{n\pi u}{l}$ and $B = \dfrac{n\pi t}{l}$, (2.2.3) becomes

(2.2.4) $f(t) = \dfrac{1}{2l}\int_{-l}^{l} f(u)du + \sum_{n=1}^{\infty} \dfrac{1}{l}\int_{-l}^{l} f(u)\cos\left\{\dfrac{\pi n}{l}(u - t)\right\}du.$

Now, imagine that $l \to \infty$, which means we no longer have a periodic function with a finite period, but rather an $f(t)$ that has a single "period" defined over the entire t-axis. In other words, $f(t)$ is now any function we wish.

What happens on the right-hand side of (2.2.4) as $l \to \infty$? The first thing we can say is that if $f(t)$ is an integrable function (the kind that engineers and physicists are generally interested in), then it bounds a finite area in the interval $-\infty < t < \infty$, and so

$$\lim_{l \to \infty} \dfrac{1}{2l}\int_{-l}^{l} f(u)du = 0.$$

Next, define

$$\lambda = \dfrac{\pi}{l},$$

and then write

$$\lambda_1 = \lambda,\ \lambda_2 = 2\lambda = \dfrac{2\pi}{l},\ \lambda_3 = 3\lambda = \dfrac{3\pi}{l},\ \ldots,\ \lambda_n = n\lambda = \dfrac{n\pi}{l},\ \ldots$$

and so on, forever. If we write

$$\Delta\lambda_n = \lambda_{n+1} - \lambda_n = \dfrac{\pi}{l},$$

and so

$$\frac{1}{l} = \frac{\Delta\lambda_n}{\pi},$$

we then have

$$f(t) = \sum_{n=1}^{\infty} \frac{\Delta\lambda_n}{\pi} \int_{-l}^{l} f(u)\cos\{\lambda_n(u-t)\}\,du.$$

Now, as $l \to \infty$, we see that $\Delta\lambda_n \to 0$; that is, $\Delta\lambda_n$ becomes ever smaller (ever more like the differential $d\lambda$), λ_n becomes the continuous variable λ, and the sum becomes an integral with respect to λ. Since the definition of λ restricts it to nonnegative values, we restrict λ to 0 to ∞, and write the $l \to \infty$ limit of $f(t)$ as

$$f(t) = \frac{1}{\pi}\int_0^{\infty} d\lambda\left[\int_{-\infty}^{\infty} f(u)\cos\{\lambda(u-t)\}\,du\right]$$

$$= \frac{1}{\pi}\int_0^{\infty} d\lambda\left[\int_{-\infty}^{\infty} f(u)\{\cos\{\lambda u\}\cos\{\lambda t\} + \sin\{\lambda u\}\sin\{\lambda t\}\}\,du\right],$$

where again we've used $\cos(A)\cos(B) + \sin(A)\sin(B) = \cos(A - B)$. Thus,

$$(2.2.5) \qquad f(t) = \frac{1}{\pi}\int_0^{\infty}\cos(\lambda t)\left\{\int_{-\infty}^{\infty} f(u)\cos(\lambda u)\,du\right\}d\lambda$$

$$+ \frac{1}{\pi}\int_0^{\infty}\sin(\lambda t)\left\{\int_{-\infty}^{\infty} f(u)\sin(\lambda u)\,du\right\}d\lambda.$$

Now suppose $f(t)$ is an even function. That is, imagine we are given $f(t)$ for $t > 0$ and then we simply *define* what happens for $t < 0$ as $f(-t) = f(t)$. Then we have

$$\int_{-\infty}^{\infty} f(u)\cos(\lambda u)\,du = 2\int_0^{\infty} f(u)\cos(\lambda u)\,du,$$

because $\cos(\lambda u)$ is even, and also

$$\int_{-\infty}^{\infty} f(u)\sin(\lambda u)\,du = 0,$$

because $\sin(\lambda u)$ is odd. Therefore,

$$f(t) = \int_{0}^{\infty} \cos(\lambda t)\left\{\frac{2}{\pi}\int_{0}^{\infty} f(u)\cos(\lambda u)\,du\right\}d\lambda.$$

That is, if $f(t)$ is even, then

(2.2.6) $$f(t) = \int_{0}^{\infty} F(\lambda)\cos(\lambda t)\,d\lambda,$$

where

(2.2.7) $$F(\lambda) = \frac{2}{\pi}\int_{0}^{\infty} f(u)\cos(\lambda u)\,du.$$

$F(\lambda)$ is called the *Fourier cosine transform* of $f(t)$. The two integrals (2.2.6) and (2.2.7) are called the *Fourier cosine transform pair*, and are usually written as $f(t) \leftrightarrow F(\lambda)$.

In the same way, if $f(t)$ is an odd function (we are given $f(t)$ for $t>0$ and simply define $f(-t) = -f(t)$ for $t<0$), then

(2.2.8) $$f(t) = \int_{0}^{\infty} F(\lambda)\sin(\lambda t)\,d\lambda,$$

where

(2.2.9) $$F(\lambda) = \frac{2}{\pi}\int_{0}^{\infty} f(u)\sin(\lambda u)\,du,$$

and now $F(\lambda)$ is called the *Fourier sine transform* of $f(t)$.

2.3 Fourier Transforms and Dirichlet's Discontinuous Integral

As a dramatic example of the use in pure mathematics of our results so far, consider the pulse-like time function $f(t)$ shown in Figure 2.3.1 where, for $t \geq 0$ and a some (arbitrary for now) positive constant,

(2.3.1)
$$f(t) = \begin{cases} 1, & 0 \leq t < a \\ 0, & t > a \end{cases}.$$

Using the cosine transform of (2.2.7) yields

$$F(\lambda) = \frac{2}{\pi} \int_0^a \cos(\lambda u)\, du = \frac{2}{\pi} \left\{ \frac{\sin(\lambda u)}{\lambda} \right\} \Bigg|_0^a,$$

or

(2.3.2)
$$F(\lambda) = \left(\frac{2}{\pi} \right) \frac{\sin(a\lambda)}{\lambda}.$$

The other integral of the cosine transform pair, in (2.2.6), says

(2.3.3)
$$f(t) = \int_0^\infty \left(\frac{2}{\pi} \right) \frac{\sin(a\lambda)}{\lambda} \cos(\lambda t)\, d\lambda = \begin{cases} 1, & 0 \leq t < a \\ 0, & t > a \end{cases}.$$

Okay, notice (if you haven't already) that (2.3.1) carefully avoided saying anything about what $f(t)$ is at $t = a$. Our result in (2.3.3), however, says

$$f(a) = \frac{2}{\pi} \int_0^\infty \frac{\sin(a\lambda)\cos(a\lambda)}{\lambda}\, d\lambda,$$

or recalling the trigonometric identity

$$\sin(a\lambda)\cos(a\lambda) = \frac{1}{2}\sin(2a\lambda),$$

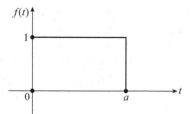

FIGURE 2.3.1. A pulse in time.

we have

$$(2.3.4) \qquad f(a) = \frac{1}{\pi} \int_0^\infty \frac{\sin(2a\lambda)}{\lambda} \, d\lambda.$$

Suppose we pick $a = \frac{1}{2}$ (remember, a has been arbitrary up to now, but we are free to give it any value we wish that is convenient). Then,

$$f\left(\frac{1}{2}\right) = \frac{1}{\pi} \int_0^\infty \frac{\sin(\lambda)}{\lambda} \, d\lambda.$$

Now, would you be terribly surprised (probably not!) if I told you that Euler had, sometime between 1776 and his death in 1783, discovered that

$$\frac{1}{\pi} \int_0^\infty \frac{\sin(\lambda)}{\lambda} \, d\lambda = \frac{1}{2}?$$

That is,

$$(2.3.5) \qquad f\left(\frac{1}{2}\right) = \frac{1}{2},$$

the *average* value of $f(t)$ for $t < a$ and for $t > a$ (with $a = \frac{1}{2}$).[9]

There has, admittedly, been an awful lot of symbol pushing so far, and so it would only be natural if you harbored this little heretical thought in the back of your mind: *Is all this stuff actually okay??* Well, let me show you an exercise, in what I call *math engineering* (recall our numerical testing of (2.1.15)), that may help convince you that in fact we haven't (yet) danced off the edge of a cliff into a fantasy world. If we combine (2.3.5) with (2.3.4), we have

$$\frac{1}{2} = \frac{1}{\pi} \int_0^\infty \frac{\sin(\lambda)}{\lambda} \, d\lambda.$$

Next, make the change of variable $\lambda = x\omega$, where x is a parameter (we can think of it as a constant for now), and ω is the new dummy integration variable. Suppose that $x > 0$. Then

$$d\lambda = x \, d\omega,$$

and so

$$d\lambda = x \, d\omega,$$

which says

$$\frac{1}{\pi} \int_0^\infty \frac{\sin(x\omega)}{x\omega} \, x \, d\omega = \frac{1}{2},$$

or

(2.3.6) $$\frac{1}{\pi} \int_0^\infty \frac{\sin(x\omega)}{\omega} \, d\omega = \frac{1}{2}, \quad x > 0.$$

In contrast, if $x < 0$, we have

$$\frac{1}{\pi} \int_0^{-\infty} \frac{\sin(x\omega)}{\omega} \, d\omega = \frac{1}{2},$$

or

(2.3.7)
$$\frac{1}{\pi} \int_{-\infty}^{0} \frac{\sin(x\omega)}{\omega} \, d\omega = -\frac{1}{2}, \quad x < 0.$$

Now, since $\frac{\sin(x\omega)}{\omega}$ is an even function of ω (because $\sin(x\omega)$ and ω are each an odd function of ω), we can write (2.3.6) and (2.3.7) as follows:

Consider

$$I(y) = \int_{0}^{\infty} e^{-y\lambda} \frac{\sin(\lambda)}{\lambda} d\lambda,$$

and so $I(0)$ is Euler's so-called *sine integral*:

$$I(0) = \int_{0}^{\infty} \frac{\sin(\lambda)}{\lambda} d\lambda.$$

Now, differentiating $I(y)$ with respect to y under the integral sign (see note 1, the comment at the end of this box, and the Appendix), we get

$$\frac{dI}{dy} = -\int_{0}^{\infty} \lambda e^{-y\lambda} \frac{\sin(\lambda)}{\lambda} d\lambda = -\int_{0}^{\infty} e^{-y\lambda} \sin(\lambda) d\lambda.$$

If you integrate this last integral by parts, *twice*, it is easy to arrive at

$$\frac{dI}{dy} = -\frac{1}{1+y^2},$$

which, in turn, is easily integrated—recall (1.3.3)—to give

(continued)

(*continued*)

$$I(y) = C - \tan^{-1}(y),$$

where C is the constant of indefinite integration. Since $I(\infty) = 0$, because the $e^{-y\lambda}$ factor in the integrand of $I(y)$ goes to zero everywhere (with the exception of the single point $\lambda = 0$) as $y \to \infty$ (because, over the entire interval of integration, $\lambda \geq 0$), we have

$$0 = C - \tan^{-1}(\infty) = C - \frac{\pi}{2},$$

and so $C = \frac{\pi}{2}$. Thus,

$$\int_0^\infty e^{-y\lambda} \frac{\sin(\lambda)}{\lambda} d\lambda = \frac{\pi}{2} - \tan^{-1}(y).$$

Now, set $y = 0$. Then, as $\tan^{-1}(0) = 0$, we immediately have

$$\int_0^\infty \frac{\sin(\lambda)}{\lambda} d\lambda = \frac{\pi}{2}.$$

Done!

Comment: The idea of differentiating under the integral sign is often called, by physicists and engineers, "Feynman's trick," after the American theoretical physicist Richard Feynman (1918–1988), who was an enthusiastic fan of the technique. Mathematicians, however, knew all about it long before Feynman was born. You can read lots more about the "trick" in my book *Inside Interesting Integrals*, Springer 2015, pp. 73–97.

(2.3.8)
$$\frac{1}{\pi} \int_{-\infty}^\infty \frac{\sin(x\omega)}{\omega} d\omega = \begin{cases} 1, & x > 0 \\ -1, & x < 0 \end{cases},$$

because, for an even function,

$$\int_{-\infty}^\infty = 2 \int_0^\infty = 2 \int_{-\infty}^0.$$

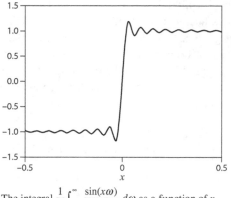

FIGURE 2.3.2. The integral $\dfrac{1}{\pi}\displaystyle\int_{-\infty}^{\infty}\dfrac{\sin(x\omega)}{\omega}\,d\omega$ as a function of x.

What happens *at* $x=0$? The integral in (2.3.8) is zero if $x=0$, because the integrand is obviously zero for every $\omega \neq 0$, and for $\omega = 0$, you can use L'Hôpital's rule to show that the integrand is still zero if $x=0$.

This is a rightfully famous result in mathematics, called *Dirichlet's discontinuous integral*,[10] and it is testable. That is, we can numerically evaluate the left-hand side of (2.3.8) for many different values of x, both negative and positive, and just look at what we get. This is easy to do today, with the ready availability of electronic computers and easy-to-use sophisticated scientific programming software. I used MATLAB, running on a quite ordinary laptop, and the code performed thousands of integrations, all to high accuracy, in less than a second. The result is shown in Figure 2.3.2 and, sure enough, we *do* see an abrupt switch from -1 to $+1$ as x passes through zero (along with the Gibbs oscillations we saw in Chapter 1 and in Figure 2.1.1 of Euler's series, oscillations that always appear in a Fourier representation of a discontinuous function).

The integral plotted in Figure 2.3.2 is well known to electrical engineers and physicists, as it occurs in theoretical studies of electric wave filters, information theory, and optics, to name just three examples. What I'll show you next is how knowing (2.3.8) can help in a pure *math* problem. Consider the scary-looking integral

$$\int_0^\infty \frac{e^{\cos(x)}\sin\{\sin(x)\}}{x}\,dx,$$

the evaluation of which is attributed to the French mathematician Augustin-Louis Cauchy (1789–1857). I discussed this integral in *Inside Interesting Integrals* as an illustration of Cauchy's theory of contour integration (see that book's pages 340 and 406), and made the claim that the integral would "otherwise be pretty darn tough [to do]." That's not so, however, as the following strictly AP-calculus solution shows.

Using Euler's fabulous formula, we can write

$$(2.3.9) \qquad e^{e^{ix}} = e^{\cos(x) + i\sin(x)} = e^{\cos(x)} e^{i\sin(x)}$$
$$= e^{\cos(x)} \left[\cos\{\sin(x)\} + i\,\sin\{\sin(x)\}\} \right].$$

Now, from the well-known power series expansion of the exponential, we have

$$e^y = \frac{1}{0!} + \frac{y}{1!} + \frac{y^2}{2!} + \frac{y^3}{3!} + \cdots + \frac{y^n}{n!} + \cdots,$$

and so, with $y = e^{ix}$, we can also write

$$e^{e^{ix}} = 1 + \frac{e^{ix}}{1!} + \frac{e^{i2x}}{2!} + \frac{e^{i3x}}{3!} + \cdots + \frac{e^{inx}}{n!} + \cdots.$$

That is,

$$(2.3.10) \quad e^{e^{ix}} = 1 + \frac{\cos(x) + i\sin(x)}{1!} + \frac{\cos(2x) + i\sin(2x)}{2!} + \cdots$$
$$+ \frac{\cos(nx) + i\sin(nx)}{n!} + \cdots,$$

or equating the imaginary parts of (2.3.9) and (2.3.10), we arrive at

$$e^{\cos(x)} \sin\{\sin(x)\} = \sum_{n=1}^{\infty} \frac{\sin(nx)}{n!}.$$

Putting this last result into Cauchy's integral, we have

$$\int_0^\infty \frac{e^{\cos(x)} \sin\{\sin(x)\}}{x} \, dx = \int_0^\infty \frac{1}{x} \sum_{n=1}^{\infty} \frac{\sin(nx)}{n!} \, dx,$$

which says

$$(2.3.11) \qquad \int_0^\infty \frac{e^{\cos(x)}\sin\{\sin(x)\}}{x}\,dx = \sum_{n=1}^\infty \frac{1}{n!}\int_0^\infty \frac{\sin(nx)}{x}\,dx,$$

if we can assume that we can reverse the order of integration and summation (which physicists and engineers routinely do and put off worrying about justifying until later, a practice that almost always causes mathematicians to roll their eyes—and with good reason, I'll admit. But being an engineer, I'll worry about that later).

You'll recognize the integral on the right in (2.3.11) as (2.3.6), where x in (2.3.11) plays the role of ω in (2.3.6), and n in (2.3.11) plays the role of x in (2.3.6). That is, for each value of n in the summation, the integral is $\frac{\pi}{2}$. Thus,

$$\int_0^\infty \frac{e^{\cos(x)}\sin\{\sin(x)\}}{x}\,dx = \frac{\pi}{2}\sum_{n=1}^\infty \frac{1}{n!}.$$

Since

$$e = \sum_{n=0}^\infty \frac{1}{n!} = 1 + \sum_{n=1}^\infty \frac{1}{n!},$$

then

$$\sum_{n=1}^\infty \frac{1}{n!} = e - 1,$$

and we have our result:

$$(2.3.12) \qquad \int_0^\infty \frac{e^{\cos(x)}\sin\{\sin(x)\}}{x}\,dx = \frac{\pi}{2}(e-1),$$

which is indeed the same answer one gets using the far more sophisticated method of contour integration.

Now, one final word about how an engineer looks at (2.3.12), when the issue of reversing the operations of integration and summation is raised. Let's do what we did for (2.1.15). That is, let's numerically evaluate the integral and see if we get something at least "reasonably close" to $\frac{\pi}{2}(e-1) = 2.69907\ldots$. MATLAB has a powerful numerical integration (or *quadrature*) command (called *integral*) which accepts three arguments: the integrand, the lower limit of integration, and the upper limit of integration. The syntax is virtually self-explanatory, and typing

$$quad(@(x)exp(cos(x)).*sin(sin(x))./x,0,1000)$$

(using 1,000 as an approximation for infinity) results in MATLAB almost instantly returning the value 2.6986 .... That's "sufficiently close" that most engineers and physicists would almost certainly cease worrying about formally justifying our integration/summation reversal.

Besides Dirichlet's integral, Euler's identity played a crucial role in the derivation of (2.3.12), and so he is almost as much the star here as is Fourier. Speaking of Euler, let me end this chapter by showing you the result of one more of his dazzling calculations. In Chapter 1, I mentioned his failure to sum the reciprocals of the positive integers *cubed*. And in the opening of this chapter, I showed you his discovery of what is probably the first Fourier series, using a geometric sum trick, along with his "fabulous formula." In 1772, Euler used that same trick to compute[11]

$$\sum_{k=1}^{\infty} \frac{1}{k^3} = \frac{2\pi^2 \ln(2) + 16 \int_0^{\pi/2} x \ln\{\sin(x)\} dx}{7}.$$

Do *that* integral and become a math superstar!

I've spent some time in this chapter on Fourier's mathematics, because it's most important that it be viewed by you as legitimate: the transform integrals, in particular, will be enormously helpful to us in finding solutions to the heat equation in Chapter 4. But, before we *solve* the heat equation, we of course have to *derive* it. That's the next chapter.

CHAPTER 3
The Heat Equation

3.1 Deriving the Heat Equation in an Infinite Mass

Imagine a block of uniform matter of infinite extent in all directions. That is, the matter fills all of space. (Only mathematicians can do this!) All of the physical properties of this block are identical at all points and unchanging with time, with one exception: the *temperature* may vary from point to point, and with time. Now, imagine further that somewhere in this infinite mass, we have a tiny closed, rectangular surface S, of dimensions Δx, Δy, Δz, as shown in Figure 3.1.1. In particular, one of the vertices of this surface is labeled P, with P's coordinates being (x_0, y_0, z_0) with respect to some origin. Indeed, let's define $u(x, y, z, t)$ as the temperature at the point (x, y, z) at time t. In addition, let's write the so-called *specific heat* of the infinite mass of matter as σ, where in the metric cgs (centimeter-gram-second) system, σ is the amount of heat energy (in calories) required to raise the temperature of 1 gram of the mass by 1 degree centigrade. For water, $\sigma = 1$.

Now, just a word on what we mean by *heat energy*. Matter can exist in many forms, from loose dirt, to solid rocks, to super-hot ionized gas (plasma), to whatever might be the state of things 10,000 miles beneath the surface of the Sun, to the incredible conditions at the center of a neutron star. For our purposes here, we will be interested in matter more like dirt and rocks than like a plasma (in which all molecular structure has been ripped apart). For us, to speak of *heat energy* is to speak of the energy of motion of the molecules of matter. We'll take this motion to be random in each direction, with no preferred direction.

That is, the molecules are *hot* (and so moving/vibrating), but they are *not* exhibiting a gross traveling motion in some specific direction. The energy of motion can be transferred from a molecule to its nearest

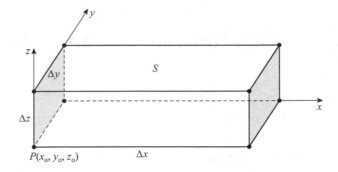

FIGURE 3.1.1. S is a closed, rectangular surface.

neighbors (like from a hot stovetop to your fingers!), and this energy-transfer mechanism is called *heat conduction*. By contrast, a gross traveling motion in which a molecule transfers energy of motion from one point to another distant point by actually making a directed journey is called *heat convection* (and that is *not* the mechanism of heat energy transfer we are considering here).

We are also not considering heat energy propagation due to *radiation*. All masses transfer heat energy to their surrounding environment, via electromagnetic radiation, according to the *Stefan-Boltzmann fourth-power law*:[1] if the surface of a mass is u, and the temperature of the surroundings is at temperature u_0, then the power radiated (energy per unit time) by the mass is proportional to $u^4 - u_0^4$ (a negative value means the mass *absorbs* energy from its environment), where the temperature is measured on the Kelvin scale (one degree on the Kelvin scale is one degree on the centigrade scale, but the zero points are different). This power law generally results in extremely difficult mathematical problems,[2] and we will *not* treat them in this book (such concerns don't arise in the Atlantic cable, in any case, which is our ultimate goal in Chapter 5).

Now, back to Figure 3.1.1. If the density of the matter in S is ρ, then the total heat energy q required to raise the temperature of the matter in S from $u = u_1$ to $u = u_2$ is

$$q = \rho \Delta x \Delta y \Delta z \sigma (u_2 - u_1) = \sigma \rho \Delta x \Delta y \Delta z \Delta u,$$

and we imagine this is at time $t = t_0$. Now, imagine that the change in temperature from u_1 to $u_2 = u_1 + \Delta u$ takes time Δt. Let's write $\frac{\partial u}{\partial t}$ as the time rate of change of the temperature, which is generally a function of where we are in the mass. If we imagine Δx, Δy, and Δz to all be very small, however, then it's reasonable to imagine $\frac{\partial u}{\partial t}$ is the same all through S and that

$$u_2 = u_1 + \frac{\partial u}{\partial t} \Delta t,$$

that is,

$$\Delta u = \frac{\partial u}{\partial t} \Delta t,$$

and so

(3.1.1) $$q = \sigma \rho \frac{\partial u}{\partial t} \Delta x \Delta y \Delta z \Delta t.$$

Okay, let's now calculate q again, but in a completely different way. Let's start by assuming there are no internal sources of heat energy in S (as there would be in radioactive matter, in which energy is continually released by natural atomic decay), and that the only way heat energy enters S is by flowing into S through the surface of S. Of course, heat energy can also flow out of S as well. We'll look at heat energy flow as a vector having three components: one along the x direction, one along the y direction, and one along the z direction, denoted by q_x, q_y, and q_z, respectively.

To calculate q_x, we look at the heat energy flowing *into* S through the left-side vertical face of S, *minus* the heat energy flowing *out* of S through the right-side vertical face of S. Before continuing, we have to make two plausible physical assumptions. (They are *plausible* because they are based on direct experimental observation and reflect what is routinely seen to be what actually happens.) First, the flow of heat energy at any point is directly proportional to the *temperature gradient* at that point, that is, to the rate of change of the temperature with respect to distance. We'll write that proportionality constant (called the *thermal conductivity*) as K. And second, heat energy flows from hot to cold (don't laugh, this is all too easy to overlook), and so heat energy flows in the direction in which temperature *drops*. That is, in the direction of a negative gradient, and so to get a positive heat energy flow from a negative gradient, we have to include a minus sign.

Thus, the heat energy flowing *into* S in the x direction, through the left-side vertical face with area $\Delta y \Delta z$ (which is perpendicular to the x direction), in time interval Δt, is

$$-K \left(\left. \frac{\partial u}{\partial x} \right|_{x=x_0} \right) (\Delta y \Delta z) \Delta t,$$

and the heat energy flowing *out* of S in the x direction, through the right-side vertical face with area $\Delta y \Delta z$, in time interval Δt, is

$$-K \left(\left. \frac{\partial u}{\partial x} \right|_{x=x_0+\Delta x} \right) (\Delta y \Delta z) \Delta t.$$

So the x-component of the heat energy flow vector is

$$q_x = \left[\left\{ -K \left(\left. \frac{\partial u}{\partial x} \right|_{x=x_0} \right) \right\} - \left\{ -K \left(\left. \frac{\partial u}{\partial x} \right|_{x=x_0+\Delta x} \right) \right\} \right] \Delta y \Delta z \Delta t,$$

or

$$q_x = K \left[\left(\left. \frac{\partial u}{\partial x} \right|_{x=x_0+\Delta x} \right) - \left(\left. \frac{\partial u}{\partial x} \right|_{x=x_0} \right) \right] \Delta y \Delta z \Delta t.$$

We can write

$$\left(\left. \frac{\partial u}{\partial x} \right|_{x=x_0+\Delta x} \right) = \left(\left. \frac{\partial u}{\partial x} \right|_{x=x_0} \right) + \Delta \left(\frac{\partial u}{\partial x} \right),$$

where $\Delta\left(\dfrac{\partial u}{\partial x}\right)$ is the amount $\dfrac{\partial u}{\partial x}$ changes as x goes from x_0 to $x_0 + \Delta x$. That change is

$$\frac{\partial}{\partial x}\left(\frac{\partial u}{\partial x}\right)\Delta x = \frac{\partial^2 u}{\partial x^2}\Delta x,$$

and so

$$\left(\frac{\partial u}{\partial x}\bigg|_{x=x_0+\Delta x}\right) = \left(\frac{\partial u}{\partial x}\bigg|_{x=x_0}\right) + \frac{\partial^2 u}{\partial x^2}\Delta x,$$

which means

$$q_x = K\frac{\partial^2 u}{\partial x^2}\Delta x \Delta y \Delta z \Delta t.$$

If you repeat this entire argument for q_y and q_z, you get

$$q_y = K\frac{\partial^2 u}{\partial y^2}\Delta x \Delta y \Delta z \Delta t$$

and

$$q_z = K\frac{\partial^2 u}{\partial z^2}\Delta x \Delta y \Delta z \Delta t.$$

As a result, because

$$q = q_x + q_y + q_z,$$

we have

$$\text{(3.1.2)} \qquad q = K\left(\frac{\partial^2 u}{\partial x^2} + \frac{\partial^2 u}{\partial y^2} + \frac{\partial^2 u}{\partial z^2}\right)\Delta x \Delta y \Delta z \Delta t.$$

If we set our two expressions, (3.1.1) and (3.1.2), for q equal, then we get

(3.1.3)
$$\frac{\partial u}{\partial t} = \frac{K}{\sigma \rho} \left(\frac{\partial^2 u}{\partial x^2} + \frac{\partial^2 u}{\partial y^2} + \frac{\partial^2 u}{\partial z^2} \right),$$

a *second-order partial differential equation* famous in mathematical physics as the *heat equation*.

It is also called the *diffusion equation*, because, if instead of heat energy of random motion we think of *molecular density* as the physical property of interest, then (3.1.3) still applies. Here's what that means. Suppose you have a glass of crystal-clear water, into which you *slowly* insert an eye-dropper filled with red ink. You do this *slowly* so as not to create any disturbing convection currents in the water. Then, you *gently* squeeze the eye-dropper, and so inject a tiny drop of ink into the water. At first, the drop is a sharply defined "point," but as time goes on, the point undergoes the following two changes: (1) it grows in radius, and (2) the surface interface boundary between the ink and the water becomes increasingly less sharply defined. After a long time, the ink "point" has completely vanished and the entire glass of water is a uniform shade of pink.

Even though the red ink molecules, individually, move at random with no preferred direction of motion, more ink molecules will move (at random) into regions of the water where the ink molecule density is low than will move from regions of low ink-molecule density into regions of high ink-molecule density. Thus, the ink spreads or diffuses throughout the water until the ink-molecule *density* is the same everywhere. Molecule *temperature* or molecule *density* (including the molecules of milk and/or sugar in your morning coffee), it's all the same to (3.1.3).

Okay, back to heat energy. We'll be particularly concerned with the one-dimensional version of (3.1.3), in which heat energy flows only in one (say, the x) direction, which is easy to arrange, as I'll discuss when we start solving actual problems. So, from now on, in this book, the words *heat equation* means

(3.1.4)
$$\frac{\partial u}{\partial t} = k \frac{\partial^2 u}{\partial x^2},$$

where the constant $k = \frac{K}{\sigma\rho}$ is called the *thermal diffusivity*.[3] This is the equation that Fourier solved, and in Chapter 4, you'll see how it's done with the mathematics of Chapter 2. But, before we do that, let me show you a little more of Fourier's genius.

3.2 Deriving the Heat Equation in a Sphere

You might think that talking of infinite masses is pretty abstract, something that can't *really* have any connection with the real world. Later, when we get to the Atlantic cable, you'll see that isn't so, but that's still a way off for us. What I'll show you, right now, is a version of the heat equation that Fourier also derived in *Analytical Theory* for a finite mass, a mass in the shape of an initially hot sphere that is losing its heat energy to a surrounding environment.

Fourier didn't elaborate on a physical example of this problem, but later, just a few decades after his death, it became the basis for a hot debate on the age of the Earth. After all, that does seem to be the origin of our planet: once, long ago, Earth was born in some cataclysmic event as a molten sphere of blinding brilliance, and then it began to cool. The question that occurred (in 1862) to William Thomson (yes, the very same fellow we'll meet again when we get to the Atlantic cable) was whether it might be possible to determine the age of the Earth by calculating, using Fourier's theory, how long it would take a molten, spherical blob of matter the size of Earth to cool from $7,000°F = 3,900°C$ (Thomson's value for a completely molten Earth—by comparison, the temperature of red-hot volcanic lava is around $1,000°C$) to its presently observed state as a planet with a solidified surface crust.[4]

More precisely, Thomson's question was: how long would it take a uniformly hot ($3,900°C$), molten Earth at time $t = 0$ to reach the observed *temperature gradient* at the Earth's surface? (He assumed an increase of $1°F$ for every 50 feet of descent.) For example, a mine shaft 1 mile deep would have a temperature at its bottom that is $106°F$ higher than the temperature at the surface. If r is the radial distance from the center of the Earth (with radius $R = 3,960$ miles), and $u(r, t)$ is the temperature at time t at distance r from the center, Thomson wanted to know the value of $t = T$ such that

$$\left.\frac{du}{dr}\right|_{r=R} = -\frac{1°F}{50 \text{ feet}} = -\frac{5/9°C}{1,524 \text{ cm}} = -3.65 \times 10^{-4} \text{ °C/cm}.$$

We'll answer Thomson's question when we get to Chapter 4.

One mid-20th-century authority[5] on the origin of the Earth described it as follows: "The usually accepted model for the newly formed earth has been a completely molten object with all metallic iron collected in a central core. Lord Kelvin [Thomson] pictured the cooling events as the formation of huge blocks of solid silicates which sank like great battleships at first, melting as they went down. Later the sinking rock reached the core's surface and solidification began at the core and proceeded outward *until the surface also solidified* [my emphasis]." There then follows these cautionary words: "This model of the earth becomes *impossible* [my emphasis] at least in its details if radioactive heating occurs. . . . Radioactive heating would . . . melt the earth again throughout except for the crust . . . with the crust sinking again." As I mentioned in the previous section, for our work in this book, we'll ignore the complications of radioactive heating.

The importance that Thomson (now Lord Kelvin) attached to the problem of a cooling Earth is shown by his words in an 1894 letter to a friend: "I would rather know the date of the *consistentior status* [the "state of greater consistency," a term for the surface crust originated a century and a half earlier by Leibniz] than of the Norman Conquest." He repeated that sentiment three years later during an 1897 address to the Victoria Institute: "The age of the earth *as an abode fitted for life* [my emphasis] is certainly a subject which largely interests mankind in general. For geology it is of vital and fundamental importance—as important as the date of the Battle of Hastings is for English history."

Kelvin's phrase "as an abode fitted for life" is most important—by that, he meant "since the appearance of the surface crust," because before that appearance, there surely would have been nothing living on the still-molten surface. Once the surface crust formed, however, it became possible for the oceans (and their abundant lifeforms) to come into existence. What Kelvin had calculated, therefore, was not the date of the *consistentior status*, but rather the total time elapsing from the creation of a uniformly hot, molten globe to a globe with today's observed surface thermal gradient. The *consistentior status* would date from some (unknown, as is clear in Kelvin's letter) intermediate time.

It *is* possible to determine that total time (the "age of the Earth") using Fourier, but the value (98 million years) that Thomson arrived at, while much larger than Buffon's estimate (see note 4), was still far too small to be consistent with either geology or with Darwin's then-new theory of evolution.[6] The predictable result of that conflict between mathematical physics, geology, and biology was a *huge* debate (one lasting decades) involving geologists, physicists, and religious scholars (who, in particular, thought the whole business silly, since Biblical study had convinced them that the world had been created by God in 4,004 BC, and so who cared about Darwin and Fourier?). And as for the existence of ancient rock strata loaded with both sea creature fossils and the clear signatures of ancient tides (ripple marks), as well as the gigantic bones of enormous, fearsome land animals (dinosaurs) that clearly had lived long before 4,004 BC, Biblical enthusiasts dismissed such things as being simply "sports of nature" created by a God with a sense of humor, for the amusement of all. Well, we won't get into any of *that*, and I'll limit myself here to simply showing you the mathematical physics of how Fourier derived the heat equation for a cooling sphere.[7]

As for why Thomson's use of Fourier for calculating the age of the Earth is so much at variance with the modern value of 5 billion (!) years, as determined by geological and astrophysical arguments (the estimated age of the Moon, which is surely less ancient than the Earth), one post-1900 suggestion was his lack of understanding concerning heating via radioactive decay. More recent views are that radioactivity, by itself, is not sufficient to explain the discrepancy, but rather it is also necessary to incorporate geological processes, such as the energy dissipation due to tectonic plate interactions (about which, again, Thomson was ignorant).

There are a number of serious computational objections one can easily raise to Thomson's very simple model of a cooling Earth. The most obvious is, I think, the value of the surface temperature gradient. Thomson himself measured different thermal gradients at various localities that ranged from 1°F for every 15 feet to 1°F for every 110 feet. And when he got to actually solving the heat equation (as we will in Chapter 4), just what numerical value should be used for the diffusivity? The value he used was simply one selected from several values he measured using various rock and sand samples from different quarries in Edinburgh, Scotland. Would that chosen value, one wonders, be at all representative of the diffusivity of the surface matter under the Arctic ice cap, or in Biloxi, Mississippi? The value used, in any case, was assumed to be a constant, rather than one dependent (in some complicated fashion) on either temperature or pressure, both of which increase (dramatically) as one descends toward the center of the Earth.

The uncertainty in all of Thomson's assumptions was pointed out in a paper[8] published the year after his death. Written by the American geologist George Becker (1847–1919), who was a senior scientist with the US Geological Survey, the paper paid proper tribute to Thomson, calling him "our great master in geophysics," but then went on to observe that the master himself had expressed concern over a number of his assumptions, with the main one being the uniformity of the initial temperature. Becker modified that assumption, and instead took the initial surface temperature to have some value less than Thomson's 3,900°C (Becker's values varied from 1,190°C to 1,453°C), which then linearly increased as one descended down to a depth of 40 miles, after which the temperature was a constant all the rest of the way to the Earth's center. The actual temperature of the inner Earth was, Becker asserted, not important—as he somewhat poetically put it, "the inner part . . . may have been originally at the temperature of ice [cold] or of the electric arc [pretty hot!]; it may conduct heat as well as silver or as ill as magnesia; in any case the influence on the outer surface would be insensible after scores of millions of years."

Becker's final (erroneous!) conclusion: a 60-million-year-old Earth with an initial surface temperature of 1,307°C was "a fair approximation to the truth and that with better data this age will not be changed by more than perhaps 5 million years." Thomson, in fact, as pointed out by Becker, fully appreciated the risks inherent in *his* assumptions, and so he placed his nominal 98 million year value for the age of the Earth within a much wider estimate of between 20 million to 400 million years.

The same year (1862) that Thomson wrote about a cooling Earth, he wrote another paper that came at the age of the Earth from a different angle. This time, he estimated the age of the Sun ("On the Age of the Sun's Heat"), a number that obviously puts an upper bound on the age of the Earth. The result (which did *not* use Fourier)—probably 100 million years and certainly no more than 500 million years—is certainly in rough agreement with his cooling-Earth analysis. Those new calculations were fatally flawed as well, however, because Thomson was unaware of the true energy source of the Sun (and of all stars, in general): thermonuclear fusion reactions occurring deep within their cores.

Stars are, literally, continuous nuclear explosions that are held to-gether (unlike a bomb) by the star's immense gravity. That under-standing came long after Thomson's death, of course, and one of the great scientific mysteries of nineteenth-century physics was that of explaining the stupendous energy output of the Sun. (Each second, 4 *million tons* of the Sun's mass is totally converted to pure energy and radiated into space.) One amusing idea was that the Sun is a giant lump of coal burning in the sky (!), but that idea was one held mostly by the uneducated: Thomson and other men of Victorian science knew it to be absurd. Much more plausible was *gravitational contrac-tion*, a process by which a giant cloud of interstellar gas would col-lapse inward due to its own internal gravity and so self-compress, thereby becoming hot to the point of glowing. This imaginative idea is due to the German physicist Hermann von Helmholtz (1821–1894), who put it forward in 1854.[9]

For our purposes here, of course, the fact that Thomson's physical models of the Earth and the Sun were wrong is beside the point. What we are going to be interested in, in Chapter 4, is given Thomson's *physi-cal* model of the Earth, how do we solve its *mathematical* description using Fourier? We start the derivation of the heat equation for a sphere by imagining that the initial temperature of a homogeneous sphere var-ies only with the radial distance r from the center. The sphere's surface is, as Fourier wrote, "exposed to air which is kept at temperature 0, and *displaced with constant velocity* [my emphasis]: it is required to deter-mine the successive states of the [sphere] during the whole time of the cooling."[10] As before, u will denote the temperature and, by the state-ment of the problem, u is a function only of the radial distance r (and, of course, of time). That is, $u = u(r, t)$.

Now, following Fourier, imagine the sphere as like an onion, built up from countless thin shells, with each shell defined by two concentric spherical surfaces of *nearly* equal radius. Consider, in particular, the two such surfaces of inner radius r_0 and outer radius $r_0 + \Delta r$. The flow of heat energy in the sphere is clearly (by symmetry) purely radial, and so the heat energy that crosses the inner surface into the shell, in time interval Δt, is

$$q = K\left(-\left.\frac{\partial u}{\partial r}\right|_{r=r_0}\right)4\pi r_0^2 \Delta t = -4K\pi r_0^2\left(\left.\frac{\partial u}{\partial r}\right|_{r=r_0}\right)\Delta t.$$

This product, factor-by-factor, is analogous to our earlier derivation. That is, the heat energy crossing the inner surface is the thermal conductivity K, times the negative of the temperature gradient at the inner surface, times the surface area that the energy crosses, times the time interval.

Heat energy is, of course, also leaving the shell by crossing the outer surface into the more remote portion of the sphere. That heat energy is given in the same way by

$$q = -4K\pi(r_0 + \Delta r)^2 \left(\left. \frac{\partial u}{\partial r} \right|_{r=r_0+\Delta r} \right) \Delta t,$$

which, if we ignore $(\Delta r)^2$ compared to Δr (remember, Δr is already small, and we will soon let it go to zero) becomes

$$q = -4K\pi(r_0^2 + 2r_0\Delta r) \left(\left. \frac{\partial u}{\partial r} \right|_{r=r_0+\Delta r} \right) \Delta t.$$

Thus, the quantity of heat energy that accumulates in the shell (and so changes the temperature of the shell) is the heat energy entering the shell minus the heat energy leaving the shell, or

$$q = -4K\pi r_0^2 \left(\left. \frac{\partial u}{\partial r} \right|_{r=r_0} \right) \Delta t - \left\{ -4K\pi(r_0^2 + 2r_0\Delta r) \left(\left. \frac{\partial u}{\partial r} \right|_{r=r_0+\Delta r} \right) \Delta t \right\}$$

$$= -4K\pi\Delta t \left[r_0^2 \left(\left. \frac{\partial u}{\partial r} \right|_{r=r_0} \right) - (r_0^2 + 2r_0\Delta r) \left(\left. \frac{\partial u}{\partial r} \right|_{r=r_0+\Delta r} \right) \right]$$

$$= 4K\pi\Delta t \left[(r_0^2 + 2r_0\Delta r) \left(\left. \frac{\partial u}{\partial r} \right|_{r=r_0+\Delta r} \right) - r_0^2 \left(\left. \frac{\partial u}{\partial r} \right|_{r=r_0} \right) \right].$$

Since

$$\left. \frac{\partial u}{\partial r} \right|_{r=r_0+\Delta r} = \left. \frac{\partial u}{\partial r} \right|_{r=r_0} + \frac{\partial}{\partial r} \left(\left. \frac{\partial u}{\partial r} \right|_{r=r_0} \right) \Delta r = \left. \frac{\partial u}{\partial r} \right|_{r=r_0} + \left. \frac{\partial^2 u}{\partial r^2} \right|_{r=r_0} \Delta r,$$

we have the heat energy accumulating in the shell, in time interval Δt, as

$$q = 4K\pi\Delta t\left[(r_0^2 + 2r_0\Delta r)\left\{\left.\frac{\partial u}{\partial r}\right|_{r=r_0} + \left.\frac{\partial^2 u}{\partial r^2}\right|_{r=r_0}\Delta r\right\} - r_0^2\left(\left.\frac{\partial u}{\partial r}\right|_{r=r_0}\right)\right],$$

which, again if we ignore $(\Delta r)^2$ compared to Δr, is

$$q = 4K\pi\Delta t\left[r_0^2\Delta r\left(\left.\frac{\partial^2 u}{\partial r^2}\right|_{r=r_0}\right) + 2r_0\Delta r\left(\left.\frac{\partial u}{\partial r}\right|_{r=r_0}\right)\right].$$

So, if we replace Δt with dt and Δr with dr, we arrive at the following for the net heat energy entering the shell:

$$(3.2.1) \qquad q = 4K\pi dtdr\left[r_0^2\left(\left.\frac{\partial^2 u}{\partial r^2}\right|_{r=r_0}\right) + 2r_0\left(\left.\frac{\partial u}{\partial r}\right|_{r=r_0}\right)\right].$$

You'll recall from our first derivation of the heat equation that the heat energy that results in the time rate of change $\frac{\partial u}{\partial t}$ is $\sigma\rho\frac{\partial u}{\partial t}$ (*volume of S*)dt, where σ is the specific heat, and ρ is the mass density. For the thin spherical shell that plays the role of S in this analysis, the volume is $4\pi r_0^2 dr$, and so the heat energy in the shell volume is

$$(3.2.2) \qquad q = \sigma\rho\frac{\partial u}{\partial t}4\pi r_0^2 drdt.$$

So, equating our two heat energy expressions, (3.2.1) and (3.2.2), we have

$$\sigma\rho\frac{\partial u}{\partial t}4\pi r_0^2 drdt = 4K\pi dtdr\left[r_0^2\left(\left.\frac{\partial^2 u}{\partial r^2}\right|_{r=r_0}\right) + 2r_0\left(\left.\frac{\partial u}{\partial r}\right|_{r=r_0}\right)\right],$$

or, since r_0 is arbitrary, we can drop the subscript and write just r:

$$\frac{\partial u}{\partial t} = \frac{K}{\sigma\rho r^2}\left[r^2\frac{\partial^2 u}{\partial r^2} + 2r\frac{\partial u}{\partial r}\right],$$

which reduces to

$$(3.2.3) \qquad \frac{\partial u}{\partial t} = k\left[\frac{\partial^2 u}{\partial r^2} + \left(\frac{2}{r}\right)\frac{\partial u}{\partial r}\right], \quad k = \frac{K}{\sigma\rho}.$$

This result looks very different from the result of our original derivation of the heat equation, but that's because in the first case, we dealt with an infinite mass filling all of space, and in this case, we are dealing with a finite, localized mass in the shape of a sphere.[11] The geometry looks different, but the physics is the same in both cases. We can see this mathematically by first recognizing that

$$\frac{\partial^2 u}{\partial r^2} + \left(\frac{2}{r}\right)\frac{\partial u}{\partial r} = \frac{1}{r}\frac{\partial^2(ur)}{\partial r^2},$$

a claim that is easily verified by actually working through the details of the right-hand side. That is,

$$\frac{1}{r}\frac{\partial^2(ur)}{\partial r^2} = \frac{1}{r}\frac{\partial}{\partial r}\left[\frac{\partial(ur)}{\partial r}\right] = \frac{1}{r}\frac{\partial}{\partial r}\left[u + r\frac{\partial u}{\partial r}\right] = \frac{1}{r}\left[\frac{\partial u}{\partial r} + \frac{\partial}{\partial r}\left(r\frac{\partial u}{\partial r}\right)\right]$$

$$= \frac{1}{r}\frac{\partial u}{\partial r} + \frac{1}{r}\frac{\partial}{\partial r}\left(r\frac{\partial u}{\partial r}\right) = \frac{1}{r}\frac{\partial u}{\partial r} + \frac{1}{r}\left[r\frac{\partial^2 u}{\partial r^2} + \frac{\partial u}{\partial r}\right] = \frac{\partial^2 u}{\partial r^2} + \left(\frac{2}{r}\right)\frac{\partial u}{\partial r},$$

as claimed. So, the heat equation for the sphere is

$$(3.2.4) \qquad \frac{\partial u}{\partial t} = k\frac{1}{r}\frac{\partial^2(ur)}{\partial r^2}.$$

Now, change the variable to $v = ur$. Then, on the left-hand side of (3.2.4), we have

$$\frac{\partial u}{\partial t} = \frac{\partial}{\partial t}\left(\frac{v}{r}\right) = \frac{r\dfrac{\partial v}{\partial t} - v\dfrac{\partial r}{\partial t}}{r^2} = \frac{1}{r}\frac{\partial v}{\partial t},$$

because r is not a function of t, and so $\frac{\partial r}{\partial t} = 0$. Thus, the heat equation in a sphere becomes

$$\frac{1}{r}\frac{\partial v}{\partial t} = k\frac{1}{r}\frac{\partial^2 v}{\partial r^2},$$

or

(3.2.5) $$\frac{\partial v}{\partial t} = k\frac{\partial^2 v}{\partial r^2}, \quad v = ur.$$

But this is just our original result for the infinite mass case, with v instead of u, and r instead of x. The sphere case *looks* different, but (3.2.5) shows that it really isn't. When, in Chapter 4, we solve the original heat equation, (3.1.4), I'll also return to the sphere version and show you how we can use the original solution to *immediately* write down the solution for a hot, cooling sphere as well.

3.3 Deriving the Heat Equation in a Very Long, Radiating Wire

While we most often associate Fourier with heat conduction, others also did important work in the subject even earlier. One of the more interesting but least well-known of those pioneers was the Dutchman Jan Ingenhousz (1730–1799). Trained as a medical doctor, his interests extended far beyond medicine; for example, he did important work on the photosynthesis of plants. A close acquaintance of Benjamin Franklin (1706–1790), a man best known today as an American revolutionary hero but also as an occasional experimenter, Ingerhousz lectured before the Royal Society of London (of which he was elected a Fellow in 1779) on Franklin's theory of positive and negative electricity. He commented, in particular, on the analogy between the conduction of heat and electricity in metals. Those observations resulted in an ingenious method for an experimental study of heat flow in metal wires. Ingenhousz obviously didn't have Fourier's heat equation on which to base his experimental work, and so what I'll show you next is a modern, theoretical presentation of what he did.

Imagine a very long, thin, metallic wire of thermal conductivity K, with the $x=0$ end of the wire held at the fixed temperature U_0. The lateral surface of the wire is not insulated, and so heat energy is lost through that surface to the surrounding medium (which we'll assume is at a temperature of zero). The far end of the wire, which for mathematical convenience we'll take as at $x=\infty$, is at the temperature of the environment (zero). Suppose that the cross-sectional area of the wire is A, and since the wire is thin, we'll further assume that we can safely ignore any temperature *variation* over any particular cross section. That is, the temperature at all points in an arbitrary cross section (at location x) is simply $u(x, t)$.

Figure 3.3.1 shows heat energy flowing through a cross section at the left, at x, into a short length (Δx) of the wire. Some of that energy H escapes from the wire through the lateral surface, and the rest continues down the wire to pass through the cross section at $x+\Delta x$. For a wire with thermal conductivity K, the heat energy per unit time passing through the cross section at x is

$$-KA\frac{du}{dx},$$

while the heat energy per unit time passing through the cross section at $x+\Delta x$ is

$$-KA\frac{d}{dx}\left(u+\frac{du}{dx}\Delta x\right) = -KA\frac{du}{dx} - KA\frac{d^2u}{dx^2}\Delta x.$$

As Figure 3.1.1 shows, and by conservation of energy,

$$-KA\frac{du}{dx} = H + \left(-KA\frac{du}{dx} - KA\frac{d^2u}{dx^2}\Delta x\right),$$

and so

$$H = KA\frac{d^2u}{dx^2}\Delta x.$$

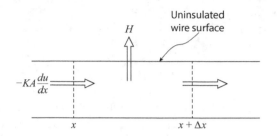

FIGURE 3.3.1. Heat energy flow along, and escape from, a thin, uninsulated metal wire.

If the circumference of the wire is c, then the area of the lateral surface is $c\Delta x$, and the heat energy loss per unit time is proportional—*if we assume Newton's law of cooling* (see note 2)—to $uc\Delta x$. If we write h as the proportionality constant (with a value dependent on the nature of the wire's environment), then

(3.3.1)
$$H = KA\frac{d^2u}{dx^2}\Delta x = huc\Delta x.$$

Since the heat energy loss is directly proportional to the temperature u, we have to modify Fourier's one-dimensional heat equation from (3.1.4) to

(3.3.2)
$$\frac{\partial u}{\partial t} = k\frac{\partial^2 u}{\partial x^2} - b^2 u,$$

where b^2 is some positive constant (the value of which we will determine in just a moment). The minus sign in (3.3.2) indicates energy is *lost* through the wire's lateral surface. We are going to defer actually solving the heat equation, in general, in its various forms, until Chapter 4, but there is one special (yet historically highly interesting) case that we can easily do, right now.

After a long time passes, the temperature of the wire, at any given point along the wire, will no longer be changing with respect to time. In that situation, the wire temperature is said to have reached its *steady state*. That is, $u(x, t) = u(x)$, and so

$$\frac{\partial u}{\partial t} = 0,$$

and (3.3.2) becomes

(3.3.3)
$$\frac{\partial^2 u}{\partial x^2} - \frac{b^2}{k} u = 0.$$

This is an easy equation to solve, using the well-known trick of *assuming* an exponential solution. That is, if we assume a steady-state solution of the form

$$u(x) = B e^{sx},$$

where B and s are arbitrary (for now) constants, then substitution of this $u(x)$ into (3.3.3) gives

$$B s^2 e^{sx} - \frac{b^2}{k} B e^{sx} = 0,$$

or

$$s^2 = \frac{b^2}{k},$$

and so

$$s = \pm \frac{b}{\sqrt{k}}.$$

Thus,

$$u(x) = B_1 e^{xb/\sqrt{k}} + B_2 e^{-xb/\sqrt{k}}.$$

Now, we certainly don't expect $u(x)$ to increase with increasing x (that is, we'd be astonished if the wire became *hotter* as we moved *away* from the input energy source at $x=0$; that just doesn't make any physical sense). To the contrary, we expect the temperature of the wire to fall as we move farther away from $x=0$. So, because by initial assumption, $u(\infty)=0$, we have $B_1=0$, and so

$$u(x) = B_2 e^{-xb/\sqrt{k}}.$$

Since $u(0)=U_0$, then $B_2=U_0$ and the steady-state temperature of the wire is

(3.3.4) $$u(x) = U_0 e^{-xb/\sqrt{k}}.$$

The big question now is, what is b?

Looking back at (3.3.1), we see that

$$\frac{d^2u}{dx^2} = \frac{hc}{KA}u.$$

Comparing this with (3.3.3), we can make the association

$$\frac{hc}{KA} = \frac{b^2}{k},$$

or

$$b^2 = \frac{khc}{KA},$$

and so, if we write

(3.3.5) $$\mu^2 = \frac{hc}{KA},$$

we have

(3.3.6) $$b^2 = k\mu^2,$$

which means $b = \mu\sqrt{k}$, and (3.3.4) becomes

(3.3.7)
$$u(x) = U_0 e^{-\mu x},$$

where μ is given by (3.3.5). This result allows us to calculate the steady-state temperature of the wire, if we know μ.

Knowledge of the value of μ requires knowledge of h (the coefficient of the "heat coupling" of the wire to the environment, more technically, called the *surface emissivity*), the cross-sectional area A of the wire, the circumference c of the wire, and the thermal conductivity K of the wire. Knowledge of all those constants may not be readily available, however (particularly that of h). We may, in fact, not even know of what metal the wire is made. There is, fortunately, an elegant alternative form for the steady-state temperature of the wire that neatly sidesteps all these concerns.

Suppose we know that at $x = l$, the temperature of the wire is T. In other words, we know (by a clever method due to Ingenhousz to be described in just a moment)

$$u(l) = T = U_0 e^{-\mu l}.$$

Thus,

$$e^{\mu l} = \frac{U_0}{T},$$

and so

(3.3.8)
$$\mu = \frac{1}{l}\ln\left(\frac{U_0}{T}\right) = \ln\left(\frac{U_0}{T}\right)^{1/l}.$$

Plugging this μ into (3.3.7) gives us

$$u(x) = U_0 e^{-x\ln\left(\frac{U_0}{T}\right)^{\frac{1}{l}}} = U_0 e^{\ln\left(\frac{U_0}{T}\right)^{-\frac{x}{l}}},$$

or

$$(3.3.9) \qquad u(x) = U_0 \left(\frac{U_0}{T} \right)^{-\frac{x}{l}}.$$

In this form of the steady-state solution, we've traded knowledge of the four constants A, K, c, and h for knowledge of the value of the single constant l, where $u(l) = T$ (where, of course, $T < U_0$).

What Ingenhousz did (circa 1780, when Fourier was not yet a teenager) was describe a clever, simple way to determine the value of l for a given U_0 and T. Suppose $U_0 = 100°C$ (easily achieved, for example, with a pan of boiling water). Placing the $x = 0$ end of the wire *the length of which he had already coated with a thin layer of beeswax* into the water, he then watched as the wire heated up along its length. The temperature of those portions of the wire near $x = 0$ eventually exceeded the melting point of beeswax (about $T = 63°C$), and the wax dripped off the wire. At points sufficiently distant from $x = 0$, however, the steady-state temperature never reached 63°C and so no melting occurred. It was then an easy task to use a ruler to measure l, the length of wire off of which the beeswax had melted. That's it![12] That was certainly a lot easier to do than trying to measure the A and the c of a thin wire, and then h for the environment. And we don't even have to know what the wire is made of, and so knowledge of K (the thermal conductivity) wasn't needed, either.

This experimental arrangement was used by Ingenhousz, in fact, to determine the relative thermal conductivities of two different (perhaps unknown) metals. Here's how. Suppose we now have *two* very long, thin, radiating wires. While they are made from different materials, they are physically identical. That is, they have the same values of A and c. We'll further assume the value of h is the same for both wires: this may not be so immediately clear, but, as long as we admit we are making an assumption here (remember Thomson's age of the Earth study, with all of its uncertain assumptions!), we can go ahead to see where it takes us. So, from (3.3.8), we find

$$\mu l = \ln\left(\frac{U_0}{T} \right),$$

and since U_0 and T are the same for both wires, then

$$\mu_1 l_1 = \mu_2 l_2,$$

or

$$\frac{l_1^2}{l_2^2} = \frac{\mu_2^2}{\mu_1^2}.$$

Using (3.3.5) for μ^2, and noting that only the thermal conductivities are different, we have

$$\frac{l_1^2}{l_2^2} = \frac{1/K_2}{1/K_1} = \frac{K_1}{K_2}.$$

In words: the thermal conductivities of the two wires are to each other as are the *squares* of the lengths of the beeswax meltings! It isn't Shakespeare, but it is nevertheless profound stuff to a mathematical physicist.

And on that semi-poetic note, we are now ready to solve Fourier's one-dimensional heat equations in all their glory, as functions of both space and time.

CHAPTER 4

Solving the Heat Equation

4.1 The Case of a Semi-Infinite Mass with a Finite Thickness

In this first example of how to solve the heat equation, we'll treat a special case, one that will both illustrate some of the tricks of the trade, as well as simultaneously solve the apparently totally unrelated case of a cooling sphere (a little extra that won't be apparent until later in this chapter.) So, let's imagine a homogeneous slab of matter that is infinite in the y and z directions (up-and-down *parallel* to the page, and in-and-out *normal* to the page, respectively), but of finite thickness in the x direction (across the page, left-to-right). Specifically, let the slab faces be at $x=0$ and $x=L$, as shown in Figure 4.1.1. We imagine, too, that the two faces are in some way kept at temperature 0, and that at time $t=0$, the temperature inside the slab is some (arbitrary) given function of x (it should be physically clear that there will be no variation of temperature in the y and z directions).

The following so-called *boundary and initial conditions* given in (4.1.1), (4.1.2), and (4.1.3), when applied to the heat equation

$$\frac{\partial u}{\partial t} = k \frac{\partial^2 u}{\partial x^2},$$

are sufficient to *uniquely* determine the temperature $u(x, t)$ anywhere in the slab at any time $t>0$. And so here is our problem: find $u(x, t)$ for $0<x<L, t>0$, subject to

(4.1.1) $u(0, t)=0,$

(4.1.2) $u(L, t)=0,$

(4.1.3) $u(x, 0)=f(x), \quad 0<x<L.$

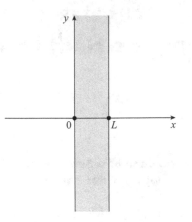

FIGURE 4.1.1. A slab of thickness L, and infinite in the y and z directions.

One last comment on the idea of an infinite slab. All that the geometry shown in Figure 4.1.1 is designed to accomplish is to force the flow of heat energy to be strictly in the x direction. We could accomplish the same goal by imagining, instead, that we have a *rod* of matter, extending from $x = 0$ to $x = L$, with its lateral surface *insulated*. If this is easier to visualize, then that's fine: it's the same physics as for the infinite-slab geometry, and a long, thin, insulated rod does conjure up the image of an electric cable, the geometry we'll discuss in Chapter 5.

What I'm going to show you in this chapter uses the sort of mathematics available to Fourier and Thomson. That is, it will be nothing but classical mathematics. As I mentioned in Chapter 1, the modern approach is to use the Laplace transform, which, while extremely elegant, does require a fair amount of preliminary development. While the classical method is longer, and in places, just a bit tricky, the math used is what a student in freshman calculus has already encountered. In addition to being historical, the classical approach is just astonishingly clever and not to be missed.

We begin with a brilliant idea dating from 1753, attributed to Daniel Bernoulli (mentioned at the start of Chapter 2), called *separation of variables*. Let's assume we can write the temperature as the product of a function of x alone and a function of t alone. That is, as

(4.1.4) $$u(x, t) = X(x)\, T(t).$$

Well, you ask, is this assumption *legitimate*? How can we justify writing (4.1.4) as the solution before we've actually solved the heat equation?[1] The answer to that is we don't know a priori that (4.1.4) is legal, but we can certainly assume it is and then see where that assumption takes us. Once we have a tentative solution, we can check it by simply seeing if it satisfies both the heat equation and the boundary/initial conditions. If the "solution" passes those tests, then it must be correct, because physically[2] we know there can be only one correct solution. The same slab, if initially heated on two different days according to the same $f(x)$ given in (4.1.3), will not cool differently on one day from how it cools on the other day. The solution, whatever it is, must be unique.[3]

A different legitimate question to ask at this point is: how did Daniel Bernoulli know to assume (4.1.4)? I don't know the answer to that with total certainty, but here's one possibility. Bernoulli was a member of a celebrated family of mathematicians. In particular, his father was Jean Bernoulli (look back at note 7 in Chapter 2), often known instead by the Anglicized form of John (or the German equivalent, Johann). Daniel's uncle (Jean's brother) was Jacques Bernoulli (1654–1705), who was often called James or Jacob. Here I'll use "John" and "Jacob." Both men were highly talented mathematicians, but they were also a jealous, combative pair.

In 1697, for example, Jacob had been trying for months, unsuccessfully, to solve the differential equation $\frac{dy}{dx} = y(x) f(x) + y^n(x) g(x)$, where $f(x)$ and $g(x)$ are two given, arbitrary functions. When John learned of his brother's failure, his spirits soared (no brotherly love here!): after all, what better way to stick a thumb in his rival's eye than to solve what Jacob couldn't, which John promptly did by making the substitution $y(x) = u(x) v(x)$? That is, John *assumed* $y(x)$ is the product of two (as yet unknown) functions, $u(x)$ and $v(x)$. Substituting this into the differential equation gives

$$v\left(\frac{du}{dx}\right) + u\left(\frac{dv}{dx}\right) = v(uf) + u(u^{n-1}v^n g),$$

and so, making the obvious associations across the equals sign, $\frac{du}{dx} = uf$ and $\frac{dv}{dx} = u^{n-1}v^n g$. Both of these equations are separable and so can be directly integrated. That is, $\frac{du}{u} = fdx$ and $\frac{dv}{v^n} = u^{n-1}gdx$, and thus (I'll let you fill in the details), $u(x) = e^{\int_0^x f(t)dt}$, and

$$v(x) = \left\{ C + (1-n)\int_0^x u^{n-1}(t)\,g(t)\,f(t)\,dt \right\}^{\frac{1}{1-n}},$$

where C is an arbitrary constant of integration.

So clever was this idea that in August 1697, the French mathematician Pierre Varignon (1654–1722) wrote to John to correctly say "In truth, there is nothing more ingenious than the solution that you give for your brother's equation, and this solution is so simple that one is surprised at how difficult the problem appeared to be: this is indeed what one calls an elegant solution." Daniel was certainly aware in 1753 of his father's then-nearly 60-year-old product trick, and so perhaps that was the inspiration for *his* product assumption. Well, okay, you say, but how did *John* know to make *his* assumption? Sorry, I'm all out of speculations!

Okay, continuing with (4.1.4), we have

$$\frac{\partial u}{\partial t} = X\frac{dT}{dt},$$

while

$$\frac{\partial^2 u}{\partial x^2} = T\frac{d^2 X}{dx^2},$$

and so the heat equation becomes

$$X\frac{dT}{dt} = kT\frac{d^2 X}{dx^2}.$$

Separating variables yields

(4.1.5)
$$\frac{1}{kT}\frac{dT}{dt} = \frac{1}{X}\frac{d^2 X}{dx^2}.$$

Now, here's the clever observation that breaks the problem wide open. The left-hand side of (4.1.5) is a function *only* of t, while the right-hand side is a function *only* of x. The only way the two sides of (4.1.5) can be equal for all x, and for all t, is if *both are equal to the same constant*. Let's call that constant α (we'll work out what it is in just a moment). Thus,

(4.1.6)
$$\frac{dT}{dt} - \alpha kT = 0$$

and

(4.1.7)
$$\frac{d^2X}{dx^2} - \alpha X = 0.$$

Furthermore, from (4.1.1) and (4.1.2), we know that (4.1.4) says

(4.1.8)
$$X(0) = 0$$

and

(4.1.9)
$$X(L) = 0.$$

If we next assume exponential solutions (a standard trick for solving ordinary linear differential equations, as are (4.1.6) and (4.1.7)), that is if we assume

$$T(t) = C_1 e^{p_1 t}$$

and

$$X(x) = C_2 e^{p_2 x},$$

where C_1, C_2, p_1, and p_2 are constants, then (4.1.6) becomes

$$C_1 p_1 e^{p_1 t} - \alpha k C_1 e^{p_1 t} = 0,$$

and (4.1.7) becomes

$$C_2 p_2^2 e^{p_2 x} - \alpha C_2 e^{p_2 x} = 0.$$

The C and the exponentials cancel away (which is precisely *why* our exponential assumption works!), and we arrive at

$$p_1 - \alpha k = 0$$

and

$$p_2^2 - \alpha = 0.$$

Thus,

$$p_1 = \alpha k$$

and

$$p_2 = \pm\sqrt{\alpha}.$$

So,

$$X(x) = C_{2a} e^{x\sqrt{\alpha}} + C_{2b} e^{-x\sqrt{\alpha}}.$$

Now, from (4.1.8), we have

$$0 = C_{2a} + C_{2b},$$

which says

$$C_{2a} = -C_{2b},$$

and so, dropping subscripts, we get

$$X(x) = C\left(e^{x\sqrt{\alpha}} - e^{-x\sqrt{\alpha}}\right).$$

From (4.1.9), we have

$$0 = C\left(e^{L\sqrt{\alpha}} - e^{-L\sqrt{\alpha}}\right)$$

or, as taking $C=0$ gives the *physically trivial* solution $X(x)=0$, we instead take

$$e^{L\sqrt{\alpha}} - e^{-L\sqrt{\alpha}} = 0.$$

This results in

(4.1.10) $e^{2L\sqrt{\alpha}} = 1.$

The obvious solution to (4.1.10) is $\alpha=0$, but this gives us the physically unacceptable solution of $X(x)=C$, a constant, a result we reject on the grounds that, surely, the slab's temperature changes with changing x. We now seem to be left with a puzzle. Where do we go from here? We have a puzzle until, that is, we remember there are a *lot* of different exponents besides zero for which

$$e^{\text{exponent}} = 1.$$

From Euler's identity, in fact, we have

$$e^{i2\pi n} = 1,$$

where n is any positive integer, $n=1, 2, 3, 4, \ldots$ (we ignore the $n=0$ case for the reason given above, and as you should convince yourself, negative values are redundant). That is,

$$2L\sqrt{\alpha} = i2\pi n,$$

or

$$\sqrt{\alpha} = in\frac{\pi}{L}, \quad n=1, 2, 3, 4, \ldots$$

as well as

$$\alpha = -\frac{\pi^2}{L^2} n^2, \quad n = 1, 2, 3, 4, \ldots.$$

Thus we have

$$X(x) = C_2 \left(e^{in\frac{\pi}{L}x} - e^{-in\frac{\pi}{L}x} \right),$$

or using Euler's identity (and absorbing a $2i$ factor into C_2),

$$X(x) = C_2 \sin\left(\frac{n\pi x}{L} \right).$$

Now, because

$$\alpha = -n^2 \frac{\pi^2}{L^2}, \quad n = 1, 2, 3, 4, \ldots,$$

and since

$$T(t) = C_1 e^{p_1 t},$$

where

$$p_1 = \alpha k = -n^2 \frac{\pi^2}{L^2} k,$$

we then have

$$T(t) = C_1 e^{-n^2 \frac{\pi^2}{L^2} k t}.$$

So, from (4.1.4) and writing the composite constant $C_1 C_2 = b_n$ (because C_1 and C_2 can, in general, be different for each choice of n), we have

(4.1.11) $\quad u(x,t) = b_n e^{-n^2 \frac{\pi^2}{L^2} kt} \sin\left(\frac{n\pi x}{L}\right), \quad n = 1, 2, 3, 4, \ldots$

Since (4.1.11) is a solution to the heat equation for each individual value of n, and as clearly a sum of those individual solutions is a solution, too, we can write the *most general* solution as

(4.1.12) $\quad u(x,t) = \sum_{n=1}^{\infty} b_n e^{-n^2 \frac{\pi^2}{L^2} kt} \sin\left(\frac{n\pi x}{L}\right), \quad 0 \le x \le L, \ t \ge 0.$

Notice that (4.1.12) does indeed satisfy (4.1.1) and (4.1.2). But what about (4.1.3)? How does (4.1.12) satisfy that requirement? As it turns out, it's easy. For $t=0$, (4.1.12) becomes

(4.1.13) $\quad u(x,0) = \sum_{n=1}^{\infty} b_n \sin\left(\frac{n\pi x}{L}\right),$

an expression that should look familiar to you.

That's because (4.1.13) is the Fourier sine series for odd periodic functions with period $2L$. In particular, for the periodic function defined as

$$u(x,0) = \begin{cases} f(x), & 0 < x < L \\ -f(x), & -L < x < 0 \end{cases},$$

where $f(x)$ is the specified temperature distribution in (4.1.3) at $t=0$. The full period includes the physically fictitious interval $-L < x < 0$, *but do we care*? No, *we don't care*, because the other half-period interval $0 < x < L$, which does have physical reality, gives the correct result of $f(x)$, and we'll confine all of our calculations to that interval. So, the b_n coefficients in (4.1.13) are the Fourier coefficients of an odd $u(x,0)$: $u(x,0)$ has a Fourier *sine series* expansion (look back at Section 2.1) where $T = 2L$.

Thus, (2.1.4) becomes

$$b_n = \frac{1}{L} \int_{-L}^{L} f(x) \sin(n\omega_0 x)\, dx = \frac{2}{L} \int_0^L f(x) \sin(n\omega_0 x) dx$$

or, since

$$\omega_0 = \frac{2\pi}{T} = \frac{2\pi}{2L} = \frac{\pi}{L},$$

(4.1.14)
$$b_n = \frac{2}{L} \int_0^L f(x) \sin\left(\frac{n\pi x}{L}\right) dx.$$

And so, at last, we have our solution; putting (4.1.14) into (4.1.12), where we use s as the dummy variable of integration in place of x (which is used elsewhere in the solution as the space variable), we have

(4.1.15)
$$u(x,t) = \frac{2}{L} \sum_{n=1}^{\infty} e^{-n^2 \frac{\pi^2}{L^2} kt} \sin\left(\frac{n\pi x}{L}\right) \int_0^L f(s) \sin\left(\frac{n\pi s}{L}\right) ds,$$

where $u(0, t) = 0$ and $u(L, t) = 0$ for $t \geq 0$, and $u(x, 0) = f(x)$ for $0 \leq x \leq L$.

The integral in (4.1.15) generally requires numerical evaluation, but if $f(x)$ is simple enough, we can do it analytically. For example, suppose $f(x) = U_0$, a constant (positive or negative). That is, the slab is initially at some uniform temperature all through its interior. Then (4.1.15) becomes

(4.1.16)
$$u(x,t) = \frac{2U_0}{L} \sum_{n=1}^{\infty} e^{-n^2 \frac{\pi^2}{L^2} kt} \sin\left(\frac{n\pi x}{L}\right) \int_0^L \sin\left(\frac{n\pi s}{L}\right) ds.$$

Since

$$\int_0^L \sin\left(\frac{n\pi s}{L}\right) ds = \left(-\frac{\cos\left(\frac{n\pi s}{L}\right)}{\frac{n\pi}{L}} \right)\Bigg|_0^L = -\frac{L}{n\pi}[\cos(n\pi) - 1]$$

$$= \begin{cases} 0, & n \text{ even} \\ \dfrac{2L}{n\pi}, & n \text{ odd} \end{cases},$$

(4.1.16) becomes

$$(4.1.17) \qquad u(x,t) = \frac{4U_0}{\pi} \sum_{n=1,3,5,\dots}^{\infty} \frac{1}{n} e^{-n^2 \frac{\pi^2}{L^2} kt} \sin\left(\frac{n\pi x}{L}\right).$$

Or since n being positive and odd means $n = 2j - 1$ as j runs through all integers from 1 to infinity, we can write

$$(4.1.18) \qquad u(x,t) = \frac{4U_0}{\pi} \sum_{j=1}^{\infty} \frac{e^{-(2j-1)^2 \frac{\pi^2}{L^2} kt}}{2j-1} \sin\left\{\frac{(2j-1)\pi x}{L}\right\}.$$

For example, suppose our infinite slab is a very large wall of concrete that is 20 cm thick, and it has been heated to a uniform temperature throughout of $U_0 = 500°C$. If each wall face is then kept at 0°C, how long will it take for the center ($x = 10$ cm) to cool down to 100°C? Taking $k = 0.005$ (in cgs units) as a typical value for the thermal diffusivity of concrete, (4.1.18) is very easy to code in any modern scientific programming language (I used MATLAB), and the answer is 15,000 seconds (4 hours and 10 minutes). If the wall is made of metal, however (say, aluminum, with $k = 0.86$), the answer is significantly less: 87 seconds (less than a minute and a half).

4.2 The Case of a Cooling Sphere

In this section, we return to the case of a sphere that is initially hot throughout its interior, according to some arbitrary function $f(r)$ of the radial distance r from the sphere's center. The radius of the sphere is R. At $t = 0$, the sphere then begins to cool because its surface is held at temperature 0°C. (Recall William Thomson's problem of a cooling Earth, discussed in Chapter 3, which we'll take up again at the end of this section.) We know from (3.2.5) that the temperature $u(r, t)$ inside the sphere is given by

$$(4.2.1) \qquad \frac{\partial v}{\partial t} = k \frac{\partial^2 v}{\partial r^2},$$

where

(4.2.2) $$v(r, t) = u(r, t)r.$$

The obvious boundary conditions on u (*not* on v, be *very* careful to note) are

(4.2.3) $$u(R, t) = 0$$

and

(4.2.4) $$u(r, 0) = f(r),$$

which, because of (4.2.2), give the boundary conditions on v as

(4.2.5) $$v(R, t) = 0$$

and

(4.2.6) $$v(r, 0) = rf(r).$$

We can also say that

(4.2.7) $$v(0, t) = 0,$$

because when $r = 0$, we have $v(0, t) = u(0, t)r = 0$.

Now, look back at the geometry we treated in the previous section: a slab with finite thickness. There we had the equation

$$\frac{\partial v}{\partial t} = k \frac{\partial^2 v}{\partial r^2}$$

with certain boundary conditions. In fact, the condition (4.1.1) looks very much like (4.2.7), the condition (4.1.2) looks very much like (4.2.5), and the condition (4.1.3) looks very much like (4.2.6). The only difference in the two problems is that, for the sphere, v appears instead of u, r appears instead of x, and $rf(r)$ appears instead of $f(x)$. The radius of the sphere

(R) plays the role of the thickness of the slab (L). So, with these substitutions, we see that we have already solved the sphere problem, and (4.1.15) becomes

$$v(r,t) = \frac{2}{R} \sum_{n=1}^{\infty} e^{-n^2 \frac{\pi^2}{R^2} kt} \sin\left(\frac{n\pi r}{R}\right) \int_0^R sf(s) \sin\left(\frac{n\pi s}{R}\right) ds.$$

Then, using (4.2.2) to write the temperature $u(r,t)$, we have our answer:

$$(4.2.8) \quad u(r,t) = \frac{2}{R} \sum_{n=1}^{\infty} e^{-n^2 \frac{\pi^2}{R^2} kt} \frac{\sin\left(\frac{n\pi r}{R}\right)}{r} \int_0^R sf(s) \sin\left(\frac{n\pi s}{R}\right) ds.$$

As we did for the slab, let's suppose the sphere is initially uniformly hot, at temperature U_0, and so $f(r) = U_0$. The integral in (4.2.8) is easily evaluated (either by parts or simply by looking it up in a table of integrals) to give

$$\int_0^R sU_0 \sin\left(\frac{n\pi s}{R}\right) ds = -\frac{U_0 R^2}{n\pi} \cos(n\pi) = \frac{U_0 R^2}{n\pi} (-1)^{n+1}, \, n = 1, 2, 3, 4, \ldots.$$

Thus,[4]

$$(4.2.9) \quad u(r,t) = \frac{2U_0 R}{\pi} \sum_{n=1}^{\infty} e^{-n^2 \frac{\pi^2}{R^2} kt} \frac{\sin\left(\frac{n\pi r}{R}\right)}{nr} (-1)^{n+1}.$$

In particular, at the center of the sphere (where, of course, $r=0$), we have

$$\lim_{r \to 0} \frac{\sin\left(\frac{n\pi r}{R}\right)}{nr} = \frac{\pi}{R},$$

and so

$$(4.2.10) \quad u(0,t) = 2U_0 \sum_{n=1}^{\infty} e^{-n^2 \frac{\pi^2}{R^2} kt} (-1)^{n+1}.$$

The series in (4.2.10) converges very quickly for all $t > 0$, but it exhibits a curious behavior *at* $t = 0$:

$$u(0,0) = 2U_0(1 - 1 + 1 - 1 + 1 - 1 + 1 - \cdots),$$

which produces a sequence of partial sums that endlessly alternates (*without converging*) between $2U_0$ and zero. But, of course, the average of $2U_0$ and zero is U_0, which is physically correct. For any $t > 0$, however, (4.2.10) *does* converge to the correct value.[5]

For example, suppose we have a sphere with a radius of 20 cm, made of iron ($k = 0.15$), initially at a uniform temperature throughout of $U_0 = 500°C$. If the surface is kept at a temperature of $0°C$, how long will it take for the temperature at the center to fall to $100°C$? Again, (4.2.10) is easy to code, and the answer is 622 seconds. If made from aluminum ($k = 0.86$), however, the time reduces to 108 seconds.

This section is titled as treating a cooling sphere (that is, with $U_0 > 0$), but our analysis can handle (with ease) the reverse case of a sphere heating up, too. As it stands, (4.2.9) is the solution for a sphere at initial temperature U_0, with its surface at $0°C$. Suppose, instead, the sphere's interior is at $0°C$ and its surface is at U_0. That is, if $U_0 > 0$, we have a sphere that will warm up with increasing time. We can treat this as the first case, with the sphere at temperature $-U_0$ (where $U_0 > 0$) and the surface at $0°C$. We then use (4.2.9) to solve for the temperature and add U_0 to the solution. This gives us the solution for a surface at $0 + U_0 = U_0$ and a sphere with an initial interior temperature of $-U_0 + U_0 = 0$. This works because a constant temperature of U_0 everywhere and everywhen is a solution to the heat equation (because all the partial derivatives with respect to t and x vanish, giving the trivial—but true!—result of $0 = 0$).

So, for a sphere with its interior initially at $0°C$ and with a surface temperature of U_0, (4.2.9) immediately tells us that the interior temperature is given by

$$(4.2.11) \quad u(r,t) = U_0 - \frac{2U_0 R}{\pi} \sum_{n=1}^{\infty} e^{-n^2 \frac{\pi^2}{R^2} kt} \frac{\sin\left(\frac{n\pi r}{R}\right)}{nr} (-1)^{n+1}.$$

In particular, at the center of the sphere, (4.2.11) reduces to

$$(4.2.12) \qquad u(0,t) = U_0 \left[1 - 2 \sum_{n=1}^{\infty} e^{-n^2 \frac{\pi^2}{R^2} kt} (-1)^{n+1} \right].$$

For example, suppose we have a granite sphere ($k=0.016$) with a radius of 15 cm, initially at 0°C all through its interior. At $t=0$, the surface temperature suddenly rises to $U_0 = 100°C$, and we ask what the temperature will be at the center at $t=2{,}000$ seconds. Coding (4.2.12) in a language like MATLAB gives us the answer: 51.6°C.

Now, to end this section, let's at last answer Thomson's question that was posed back in Chapter 3. Given a sphere of radius R, at uniform temperature U_0 at time $t=0$, at what time $t=T$ is the thermal gradient *at the surface* equal to a specified value if the surface is held at temperature 0°C? The thermal gradient at radial distance r and at time t is easily calculated by differentiating (4.2.9) with respect to r:

$$\frac{\partial u}{\partial r} = \frac{2U_0 R}{\pi} \sum_{n=1}^{\infty} e^{-n^2 \frac{\pi^2}{R^2} kt} (-1)^{n+1} \frac{d}{dr} \left[\frac{\sin\left(\frac{n\pi r}{R}\right)}{nr} \right],$$

which, at the surface ($r=R$), is[6]

$$(4.2.13) \qquad \left. \frac{\partial u}{\partial r} \right|_{r=R} = -\frac{2U_0}{R} \sum_{n=1}^{\infty} e^{-n^2 \frac{\pi^2}{R^2} kt}.$$

At $t=0$, the thermal gradient at the surface is minus infinity (*at* $t=0$, every term in the sum is one, and so the sum blows up[7]) and then, as time increases, the gradient monotonically decreases toward zero. For Thomson's problem, we have $U_0 = 3{,}900°C$, $R = 3{,}960$ miles, and $k=0.01178$, and it is easy to numerically evaluate (4.2.13) for these values and to simply try various values of t until we get Thomson's observed thermal gradient (which you'll recall from Section 3.2 was $-3.65 \times 10^{-4}°C/cm$). Here's what (4.2.13) tells us:

t (seconds)	gradient (°C/cm)
2×10^{15}	-4.47×10^{-4}
3×10^{15}	-3.64×10^{-4}
4×10^{15}	-3.14×10^{-4}

A value for t equal to 3×10^{15} seconds appears to be just about right, and converting to years (ignoring any high-order correction for leap years[8]), we get $T = 95$ million years, in good agreement with Thomson's conclusion.

4.3 The Case of a Semi-Infinite Mass with Infinite Thickness

Now, for the next detailed analysis of this chapter, the one that will (as you'll soon see) serve as the model for William Thomson's analysis of the Atlantic cable, we'll solve the heat equation for a semi-infinite solid. This may seem to be, initially, much like our first analysis for a slab with finite thickness. It may seem, at first thought, that all we need do is let $L \rightarrow \infty$ in (4.1.15). If you do that, however, the expression for $u(x, t)$ appears to vanish everywhere and everywhen, which makes no physical sense. To arrive at a realistic answer to the $L \rightarrow \infty$ case will require a new approach. That's because the $L < \infty$ case had *two* faces to the slab, and both were given as being at temperature 0°C (recall (4.1.1) and (4.1.2)). In our new analysis, we have just one face, at $x = 0$ (there is no face at $x = \infty$), and now we'll also assume the temperature at $x = 0$ is some constant (arbitrary) value (call it U_0) that isn't required to be zero. This assumption of $U_0 \neq 0$ is a complication that, as you'll see, will require some new tricks. So, because this new analysis is for a situation that is the foundation for Thomson's electric cable analysis, I'm going to show you *in detail* one way to arrive at the solution.

We start, of course, with the heat equation:

$$\frac{\partial u}{\partial t} = k \frac{\partial^2 u}{\partial x^2},$$

with the assumption that initially, the semi-infinite body is everywhere at temperature zero (except at $x=0$). That is, the initial condition is

(4.3.1) $$u(x, 0) = 0, \quad x > 0,$$

and the boundary condition is

(4.3.2) $$u(0, t) = U_0, \quad t \geq 0.$$

Also, whatever $u(x, t)$ may be, we'll physically require that

(4.3.3) $$\lim_{x \to \infty} u(x, t) = 0, \quad t \geq 0.$$

This is a statement of the fact that, as we become ever more distant from the $x = 0$ face, the temperature can't change (much) from the initial value of zero, in response to the U_0 input at $x = 0$.

As I mentioned earlier, it's (4.3.2) that is the origin of the difficulty of a direct assault on the heat equation. That difficulty would vanish if we could arrange to have a zero boundary condition at $x = 0$ (mathematicians call this the *homogeneous* case), and we can do that with the clever idea of *dimensionless normalization*. That is, suppose we change variables as follows:

(4.3.4) $$\hat{u} = \frac{U_0 - u}{U_0} = 1 - \frac{u}{U_0}.$$

Also, if we imagine L to be some fundamental unit of distance (since we have been doing all of our numerical work in the metric cgs system, let's take $L = 1$ centimeter), then define

(4.3.5) $$\hat{x} = \frac{x}{L}.$$

Notice that both $\hat{u}$ and $\hat{x}$ are pure numbers, with no dimensions. Finally, define

(4.3.6) $$\hat{t} = k \frac{t}{L^2},$$

and you'll notice that $\hat{t}$ is also dimensionless, because k has the dimensions cm²/second.

What makes these variable changes useful is how they impact both the heat equation and the boundary conditions. The heat equation transforms as follows:

$$\frac{\partial u}{\partial t} = \frac{\partial\{U_0(1-\hat{u})\}}{\partial\left\{\dfrac{L^2}{k}\hat{t}\right\}} = -\frac{U_0 k}{L^2}\frac{\partial \hat{u}}{\partial \hat{t}},$$

and

$$k\frac{\partial^2 u}{\partial x^2} = k\frac{\partial}{\partial x}\left\{\frac{\partial u}{\partial x}\right\} = k\frac{\partial}{\partial\{\hat{x}L\}}\left\{\frac{\partial\{U_0(1-\hat{u})\}}{\partial\{\hat{x}L\}}\right\} = -\frac{U_0 k}{L^2}\frac{\partial^2 \hat{u}}{\partial \hat{x}^2}.$$

Thus, putting these results into the heat equation, we get

$$-\frac{U_0 k}{L^2}\frac{\partial \hat{u}}{\partial \hat{t}} = -\frac{U_0 k}{L^2}\frac{\partial^2 \hat{u}}{\partial \hat{x}^2}$$

or, making the obvious cancellation,

$$(4.3.7) \qquad\qquad \frac{\partial \hat{u}}{\partial \hat{t}} = \frac{\partial^2 \hat{u}}{\partial \hat{x}^2}.$$

This is just the heat equation again (but in normalized dimensionless form) with a k numerically equal to 1. That's nice, but what is *really* nice is what has happened to the boundary conditions. That is, we now have

$$(4.3.8) \qquad\qquad \hat{u}(\hat{x},0) = 1, \quad \hat{x} > 0$$

and

$$(4.3.9) \qquad\qquad \hat{u}(0,\hat{t}) = 0, \quad \hat{t} > 0.$$

In particular, compare (4.3.9) with (4.3.2).

Now, at last, we are all set to solve (4.3.7), subject to the conditions of (4.3.1) and (4.3.2). We start as we did for the finite-thickness slab by using Bernoulli's trick of assuming we can separate variables and so write

$$(4.3.10) \qquad \hat{u}(\hat{x}, \hat{t}) = X(\hat{x}) \, T(\hat{t}).$$

Then, substituting back into the dimensionless heat equation, we get

$$\frac{1}{T} \frac{dT}{d\hat{t}} = \frac{1}{X} \frac{d^2 X}{d\hat{x}^2}$$

and this can be true only if both sides are equal to the same constant. Let's take advantage of what we learned in the slab analysis, and write this constant as a *negative* quantity (look back to just after (4.1.10) and to the nature of α). Thus, writing our constant as $-\lambda^2$, we have

$$(4.3.11) \qquad \frac{dT}{d\hat{t}} + \lambda^2 \, T = 0$$

and

$$(4.3.12) \qquad \frac{d^2 X}{d\hat{x}^2} + \lambda^2 X = 0.$$

The general solutions to (4.3.11) and (4.3.12) are easily verified (by direct substitution) to be

$$X(\hat{x}) = C_1 \cos(\lambda \hat{x}) + C_2 \sin(\lambda \hat{x})$$

and

$$T(\hat{t}) = C_3 e^{-\lambda^2 \hat{t}},$$

where C_1, C_2, and C_3 are arbitrary constants (I'll say more about the nature of λ in just a moment). Thus,

$$\hat{u}(\hat{x}, \hat{t}) = C_3 e^{-\lambda^2 \hat{t}} [C_1 \cos(\lambda \hat{x}) + C_2 \sin(\lambda \hat{x})],$$

or combining constants in the obvious way to arrive at the constants A and B,

(4.3.13) $$\hat{u}(\hat{x}, \hat{t}) = Ae^{-\lambda^2 \hat{t}}\cos(\lambda\hat{x}) + Be^{-\lambda^2 \hat{t}}\sin(\lambda\hat{x}).$$

Now, for (4.3.13) to satisfy (4.3.9), we must have

$$\hat{u}(0, \hat{t}) = 0 = Ae^{-\lambda^2 \hat{t}},$$

which means $A = 0$. Thus, (4.3.13) reduces to

(4.3.14) $$\hat{u}(\hat{x}, \hat{t}) = Be^{-\lambda^2 \hat{t}}\sin(\lambda\hat{x}).$$

We now make a new observation: λ is an arbitrary constant that—unlike the constant α in Section 4.1 that takes on discrete values—can *continuously* assume values. This is because in Section 4.1, our mass had two ends, but now we have just one. So, (4.3.14) holds for all possible choices for λ (since λ is squared, this means $0 < \lambda^2 < \infty$, and so using a negative value for λ adds nothing beyond what using a positive λ does). Thus, if we add terms like (4.3.14) for all possible λ—that is, if we integrate (remember, λ varies continuously) over all possible nonnegative λ—we will have a *general* solution. Furthermore, for each choice of λ, B could be a different constant as well (the word *constant* simply means that λ and B do not depend on either $\hat{x}$ or $\hat{t}$). That is, $B = B(\lambda)$. So, the most general solution is

(4.3.15) $$\hat{u}(\hat{x}, \hat{t}) = \int_0^\infty B(\lambda)e^{-\lambda^2 \hat{t}}\sin(\lambda\hat{x})d\lambda.$$

We can find $B(\lambda)$ by using (4.3.8), which gives us

(4.3.16) $$\hat{u}(\hat{x}, 0) = 1 = \int_0^\infty B(\lambda)\sin(\lambda\hat{x})d\lambda.$$

This immediately prompts an obvious question: how do we solve (4.3.16) for $B(\lambda)$, which is inside an integral? In fact, we already have, way back in Chapter 2 when we developed the Fourier sine transform.

Looking back at (2.2.8) and (2.2.9), we see that we can write if

$$f(\hat{x}) = \int_0^\infty F(\lambda) \sin(\lambda\hat{x})d\lambda,$$

then

$$F(\lambda) = \frac{2}{\pi} \int_0^\infty f(\hat{x}) \sin(\lambda\hat{x}) \, d\hat{x}.$$

Looking now at (4.3.16), we see that $B(\lambda)$ plays the role of $F(\lambda)$, and that $f(\hat{x}) = 1$, and so, just like that, we have our solution for $B(\lambda)$:

$$B(\lambda) = \frac{2}{\pi} \int_0^\infty \sin(\lambda\hat{x}) \, d\hat{x} = \frac{2}{\pi} \int_0^\infty \sin(\lambda s) \, ds,$$

where I've changed the dummy variable of integration from $\hat{x}$ to s to avoid confusion with the use of $\hat{x}$ as a *non*-dummy variable in (4.3.15). Thus, (4.3.15) becomes

$$\hat{u}(\hat{x}, \hat{t}) = \int_0^\infty \left\{ \frac{2}{\pi} \int_0^\infty \sin(\lambda s) ds \right\} \sin(\lambda\hat{x}) \, e^{-\lambda^2\hat{t}} \, d\lambda,$$

or reversing the order of integration,

$$(4.3.17) \qquad \hat{u}(\hat{x}, \hat{t}) = \frac{2}{\pi} \int_0^\infty \left\{ \int_0^\infty \sin(\lambda s) \sin(\lambda\hat{x}) e^{-\lambda^2\hat{t}} \, d\lambda \right\} ds.$$

Recalling the trigonometric identity

$$\sin(\alpha) \sin(\beta) = \frac{1}{2} [\cos(\alpha - \beta) - \cos(\alpha + \beta)],$$

we can write (4.3.17) as

$$(4.3.18) \quad \hat{u}(\hat{x}, \hat{t}) = \frac{1}{\pi} \int_0^\infty \left\{ \int_0^\infty \cos\{\lambda(s - \hat{x})\} e^{-\lambda^2\hat{t}} d\lambda \right\} ds$$

$$- \frac{1}{\pi} \int_0^\infty \left\{ \int_0^\infty \cos\{\lambda(s + \hat{x})\} e^{-\lambda^2\hat{t}} d\lambda \right\} ds.$$

The inner integrals in (4.3.18) can be found by simply using a good math table:[9]

$$\int_0^\infty e^{-ap^2}\cos(bp)\,dp = \frac{1}{2}\sqrt{\frac{\pi}{a}}\,e^{-b^2/4a}.$$

So with $p = \lambda$, $a = \hat{t}$, and $b = s \pm \hat{x}$, we have

$$\int_0^\infty \cos\{\lambda(s-\hat{x})\}e^{-\lambda^2\hat{t}}\,d\lambda = \frac{1}{2}\sqrt{\frac{\pi}{\hat{t}}}\,e^{-(s-\hat{x})^2/4\hat{t}}$$

and

$$\int_0^\infty \cos\{\lambda(s+\hat{x})\}e^{-\lambda^2\hat{t}}\,d\lambda = \frac{1}{2}\sqrt{\frac{\pi}{\hat{t}}}\,e^{-(s+\hat{x})^2/4\hat{t}},$$

which gives

(4.3.19) $$\hat{u}(\hat{x},\hat{t}) = \frac{1}{2\sqrt{\pi\hat{t}}}\left[\int_0^\infty e^{-(s-\hat{x})^2/4\hat{t}}\,ds - \int_0^\infty e^{-(s+\hat{x})^2/4\hat{t}}\,ds\right].$$

Now, change the variable in the two integrals of (4.3.19) to

$$y = \frac{s \pm \hat{x}}{2\sqrt{\hat{t}}},$$

where we use the minus sign in the first integral and the plus sign in the second integral. Then,

$$\hat{u}(\hat{x},\hat{t}) = \frac{1}{2\sqrt{\pi\hat{t}}}\left[\int_{-\hat{x}/2\sqrt{\hat{t}}}^\infty e^{-y^2}2\sqrt{\hat{t}}\,dy - \int_{\hat{x}/2\sqrt{\hat{t}}}^\infty e^{-y^2}2\sqrt{\hat{t}}\,dy\right]$$

$$= \frac{1}{\sqrt{\pi}}\int_{-\hat{x}/2\sqrt{\hat{t}}}^{\hat{x}/2\sqrt{\hat{t}}} e^{-y^2}\,dy,$$

or, as e^{-y^2} is an even function about $y = 0$,

$$(4.3.20) \qquad \hat{u}(\hat{x}, \hat{t}) = \frac{2}{\sqrt{\pi}} \int_0^{\hat{x}/2\sqrt{\hat{t}}} e^{-y^2} \, dy.$$

The integral in (4.3.20) appears in mathematical analyses so often that it has been evaluated in math tables, is available as a callable function in most scientific programming languages, and has been given its own name: the *error function*,[10] written as erf. Specifically,

$$(4.3.21) \qquad \mathrm{erf}(z) = \frac{2}{\sqrt{\pi}} \int_0^z e^{-y^2} \, dy,$$

where the $\frac{2}{\sqrt{\pi}}$ has been included, so that $\mathrm{erf}(\infty) = 1$. Clearly, $\mathrm{erf}(0) = 0$. This range of values[11] is attractive, because the error function plays a very big role in probability theory (and all probabilities are, of course, from 0 to 1), and while our work in this book has nothing to do with probability, that's why erf(z) is defined as it is.

In any case, we now have

$$(4.3.22) \qquad \hat{u}(\hat{x}, \hat{t}) = \mathrm{erf}\left(\frac{\hat{x}}{2\sqrt{\hat{t}}} \right).$$

This is the temperature—*in normalized form*—of the semi-infinite mass, and to get the actual temperature, we simply substitute what $\hat{u}$, $\hat{x}$, and $\hat{t}$ are in terms of u, x, and t, using (4.3.4), (4.3.5), and (4.3.6), respectively. So, using

$$1 - \frac{u(x,t)}{U_0} = \mathrm{erf}\left(\frac{\dfrac{x}{L}}{2\sqrt{k\dfrac{t}{L^2}}} \right) = \mathrm{erf}\left(\frac{x}{2\sqrt{kt}} \right),$$

we have, at last,

$$(4.3.23) \qquad u(x,t) = U_0\left[1 - \mathrm{erf}\left(\frac{x}{2\sqrt{kt}} \right) \right].$$

This is seen, by inspection, to satisfy the boundary conditions in (4.3.1) and (4.3.2), as well as the limiting requirement of (4.3.3). We'll do more with (4.3.23) in Chapter 5, when we discuss William Thomson's application of it to the Atlantic cable.

4.4 The Case of a Circular Ring

In his *Analytical Theory*, Fourier treated a number of different mass geometries, and the first three sections of this chapter have been specifically written to lead up to the electrical case of the Atlantic cable. In the final two sections of this chapter, I want to show you two more geometries that Fourier discussed. Although they have nothing to do with the Atlantic cable, they introduce yet new twists to solving the heat equation. At first glance, they might seem to result in problems far more difficult than the ones we've already treated, but in fact, we'll find that all of our earlier tricks still work and that the new geometries remain within our reach.

So, for my first example of this, imagine a thin wire of length $2L$ in the shape of a closed loop. That is, a *ring*, as illustrated in Figure 4.4.1. The surface of the wire is imagined to be insulated, and so all heat-energy flow is strictly along the length of the wire. This means that if we measure the spatial variable x along the curve of the wire,[12] the one-dimensional heat equation (3.1.4) still holds true:

$$(4.4.1) \qquad \frac{\partial u}{\partial t} = k \frac{\partial^2 u}{\partial t^2}.$$

What's new about the ring geometry is the nature of the boundary conditions. In fact, as you can see from Figure 4.4.1, even though the mass is finite, there are *no physical boundaries*! The ring just goes round and round, and so, even while of finite mass, it has no end. This is the first new "twist" (pun intended) that I mentioned earlier.

To start our analysis, let's imagine that the wire has been given some initial ($t = 0$) temperature distribution

$$(4.4.2) \qquad u(x, 0) = f(x).$$

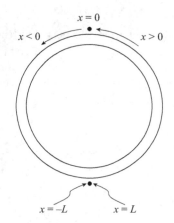

FIGURE 4.4.1. An insulated, hot wire as a circular loop.

The boundary conditions our solution must satisfy come from the physics and geometry of the closed loop formed by the wire. First, since $x=-L$ and $x=L$ are the same point, we obviously have

(4.4.3)
$$u(L, t) = u(-L, t), \quad t \geq 0.$$

Second, again because they are the same point, the thermal gradient at $x=L$ must equal the thermal gradient at $x=-L$. That is,

(4.4.4)
$$\left.\frac{\partial u}{\partial x}\right|_{x=-L} = \left.\frac{\partial u}{\partial x}\right|_{x=-L}, \quad t \geq 0.$$

We now proceed as in the previous sections, by assuming we can separate variables, which leads to

$$\frac{dT}{dt} + \lambda^2 kT = 0$$

and

$$\frac{d^2 X}{dx^2} + \lambda^2 X = 0.$$

Have you noticed, right from the very first time we invoked separation of variables, that it has been a completely arbitrary decision on which side of the separated equation we placed the thermal diffusivity k? That is, the mathematics would have been just as correct if I had, instead of the last two equations, written

$$\frac{dT}{dt} + \lambda^2 T = 0$$

and

$$\frac{d^2X}{dx^2} + \frac{\lambda^2}{k} X = 0.$$

Nature, after all, doesn't care on which side I put k—but then what about the final solution, which is a *physical quantity* (temperature), and Nature cares a *lot* about that? This is a point that is typically glossed over in textbooks (or, more commonly, never even mentioned). So, here's a little exercise for you. Once we're done with the analysis here, go back and do it again using the alternative pair of equations I just wrote, and show that you do, indeed, get the same final result for the temperature. *Nature will be pleased.*

So, as we've argued before,

$$(4.4.5) \qquad u(x,t) = e^{-\lambda^2 kt}[A\cos(\lambda x) + B\sin(\lambda x)],$$

where A, B, and λ are constants. Now, from (4.4.3), we have

$$e^{-\lambda^2 kt}[A\cos(\lambda L) + B\sin(\lambda L)] = e^{-\lambda^2 kt}[A\cos(\lambda L) - B\sin(\lambda L)],$$

which reduces to

$$2B\sin(\lambda L) = 0.$$

This means either $B=0$, or $\lambda L=n\pi$, where n is any integer. We'll decide which condition to choose in just a moment.

Also, since

$$\frac{\partial u}{\partial x} = e^{-\lambda^2 kt}[-A\lambda\sin(\lambda x) + B\lambda\cos(\lambda x)],$$

from (4.4.4), we have

$$e^{-\lambda^2 kt}[-A\lambda\sin(\lambda L) + B\lambda\cos(\lambda L)] = e^{-\lambda^2 kt}[A\lambda\sin(\lambda L) + B\lambda\cos(\lambda L)],$$

which reduces to

$$2\,A\sin(\lambda L) = 0.$$

This means either $A=0$ or again, $\lambda L=n\pi$, where n is any integer.

So, our choice is clear—we satisfy *both* boundary conditions (4.4.3) and (4.4.4) at once by choosing

(4.4.6) $$\lambda = \frac{n\pi}{L}, \quad n = 0, 1, 2, 3, \ldots.$$

Putting (4.4.6) into (4.4.5) we find that for any integer n, a solution to the heat equation in a ring is

$$u_n(x,t) = e^{-\left(\frac{n\pi}{L}\right)^2 kt}\left[A_n\cos\left(\frac{n\pi}{L}x\right) + B_n\sin\left(\frac{n\pi}{L}x\right)\right], \quad n = 0, 1, 2, 3, \ldots.$$

The most general solution is simply the sum over all possible n of the $u_n(x, t)$:

$$u(x,t) = \sum_{n=0}^{\infty} u_n(x,t) = \sum_{n=0}^{\infty} e^{-\left(\frac{n\pi}{L}\right)^2 kt}\left[A_n\cos\left(\frac{n\pi}{L}x\right) + B_n\sin\left(\frac{n\pi}{L}x\right)\right],$$

or pulling the $n=0$ term (for which the exponential factor is unity, for all t) outside the sum:

$$(4.4.7) \quad u(x,t) = A_0 + \sum_{n=1}^{\infty} e^{-\left(\frac{n\pi}{L}\right)^2 kt} \left[A_n \cos\left(\frac{n\pi}{L}x\right) + B_n \sin\left(\frac{n\pi}{L}x\right) \right].$$

The obvious question now is, what are the A_i and the B_i in (4.4.7)? That question is answered by applying the initial condition of (4.4.2) to (4.4.7). Setting $t=0$ in (4.4.7), we have

$$(4.4.8) \quad u(x,0) = f(x) = A_0 + \sum_{n=1}^{\infty} \left[A_n \cos\left(\frac{n\pi}{L}x\right) + B_n \sin\left(\frac{n\pi}{L}x\right) \right],$$

and this should immediately look familiar to you (look back at (2.1.1), with k replaced by n, t replaced by x, and $\omega_0 = \frac{\pi}{L}$, which goes with a periodic function with period $2L$). That is, the A_i and the B_i are the Fourier coefficients of $f(x)$, with the proviso that $A_0 = \frac{1}{2}a_0$.

As a specific example, suppose the initial temperature distribution in the ring in Figure 4.4.1 is

$$(4.4.9) \qquad u(x,0) = f(x) = U_0 \left(\frac{x}{L}\right)^2, \quad -L \le x \le L.$$

That is, the initial temperature at $x=0$ is zero and, as we move away from $x=0$ (in either direction) the initial temperature increases quadratically until, at $x=\pm L$, the temperature is U_0 (where U_0 is a specified value). Since the ring is a closed system because of its insulated surface, no heat energy is lost as time increases. Instead, the heat energy *redistributes*. That is, the hotter portions of the ring will cool, and the cooler portions of the ring will become hotter, and this continues until the entire ring reaches a uniform temperature somewhere between zero and U_0. That uniform temperature will in fact be, from (4.4.3),

$$\lim_{t \to \infty} u(x, t) = A_0.$$

To find $u(x, t)$, we need to find the Fourier series expansion for $u(x, 0) = f(x)$, as given in (4.4.9). We know $B_n = 0$ for all n because $f(x)$ is even, and we know (with $T = 2L$)

$$a_0 = 2A_0 = \frac{2}{2L} \int_{-L}^{L} U_0 \left(\frac{x}{L} \right)^2 dx = \frac{2}{3} U_0,$$

and so

(4.4.10)
$$A_0 = \frac{1}{3} U_0.$$

This is the eventual uniform temperature of the ring as $t \to \infty$. For example, if we initially heat the ring so that $U_0 = 300°C$ at the point $x = \pm L$, the ring will then (after a "long" time) reach a uniform temperature everywhere of $100°C$. That is, the point $x = 0$ will heat up from $0°C$ to $100°C$, while point $x = \pm L$ will cool down from $300°C$ to $100°C$.

Continuing with the calculation of the Fourier series of $f(x)$, we have (with (2.1.3) as a guide)

$$A_n = \frac{2}{2L} \int_{-L}^{L} U_0 \left(\frac{x}{L} \right)^2 \cos\left(\frac{n\pi}{L} x \right) dx = \frac{2U_0}{L^3} \int_{0}^{L} x^2 \cos\left(\frac{n\pi}{L} x \right) dx,$$

or changing the variable to $y = \dfrac{\pi}{L} x$,

$$A_n = \frac{2U_0}{L^3} \int_{0}^{\pi} \left(\frac{L}{\pi} \right)^2 y^2 \cos(ny) \frac{L}{\pi} dy = \frac{2U_0}{\pi^3} \int_{0}^{\pi} y^2 \cos(ny) dy.$$

Recall that we did this integral during the derivation of (2.1.6):

$$A_n = \frac{2U_0}{\pi^3}\left[\frac{2y}{n^2}\cos(ny) + \left(\frac{y^2}{n} - \frac{2}{n^3}\right)\sin(ny)\right]\Bigg|_0^\pi = \frac{4U_0}{\pi^2 n^2}(-1)^n.$$

So

$$u(x,t) = \frac{1}{3}U_0 + \sum_{n=1}^\infty e^{-\left(\frac{n\pi}{L}\right)^2 kt}\frac{4U_0}{\pi^2 n^2}(-1)^n\cos\left(\frac{n\pi}{L}x\right),$$

or

$$(4.4.11) \quad u(x,t) = U_0\left[\frac{1}{3} + \frac{4}{\pi^2}\sum_{n=1}^\infty \frac{e^{-\left(\frac{n\pi}{L}\right)^2 kt}}{n^2}(-1)^n\cos\left(\frac{n\pi}{L}x\right)\right].$$

Setting $x=0$ in (4.4.11), and then $x=\pm L$, Figure 4.4.2 is a semi-log plot of $u(0, t)$ and $u(\pm L, t)$, for $U_0 = 300°C$, for a thin ring of size $L = 100$ cm (a ring circumference of 200 cm) made of iron ($k = 0.15$).

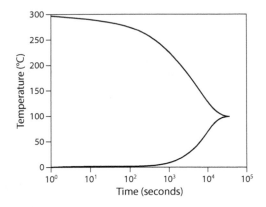

FIGURE 4.4.2. $u(0, t)$ is the bottom curve, and $u(\pm L, t)$ is the top curve.

4.5 The Case of an Insulated Sphere

For the final section of this chapter, I'll show you yet another special case treated by Fourier in his *Analytical Theory*: a sphere of radius R with an insulated surface. So, just as with the ring geometry, the initial heat energy is forever trapped inside the mass and the sphere will, as time goes by, simply redistribute that energy until the entire sphere is at a uniform temperature. Since this particular analysis plays no role in the Atlantic cable discussion of Chapter 5, I'll keep these remarks brief. But Fourier's analysis will, as did the ring analysis, illustrate yet another twist to solving the heat equation, as well as show you the sensitivity of the nature of a solution to the nature of the boundary conditions. In this case, in fact, you'll see that the solution is *not* in the form of a Fourier series!

From (4.2.1), we have the heat equation for a sphere:

$$(4.5.1) \qquad \frac{\partial v}{\partial t} = k \frac{\partial^2 v}{\partial r^2},$$

where, if $u(r, t)$ is the temperature (what we are actually interested in), then $v(r, t) = ru(r, t)$. In particular, at the sphere's center, where (by definition!) $r = 0$, we see that

$$(4.5.2) \qquad v(0, t) = 0.$$

Also, because the surface (at $r = R$) is insulated, we know that there is no heat energy flow there, and so the temperature gradient at the surface must be zero. That is, for all $t \geq 0$,

$$(4.5.3) \qquad \left. \frac{\partial u}{\partial r} \right|_{r=R} = 0.$$

Using our now-familiar separation of variables trick, let's write $v(r, t) = g(r)h(t)$, which quickly gives us the equations

$$\frac{dh}{dt} + \lambda^2 kh = 0$$

and

$$\frac{d^2g}{dr^2} + \lambda^2 g = 0,$$

where λ is an arbitrary (at this point) constant. The general solutions to these equations are easily verified to be

$$g(r) = C_1 \cos(\lambda r) + C_2 \sin(\lambda r)$$

and

$$h(t) = C_3 e^{-\lambda^2 kt},$$

where C_1, C_2, and C_3 are constants. Thus, the general solution for $v(r, t)$ is

(4.5.4) $$v(r, t) = e^{-\lambda^2 kt}[A \cos(\lambda r) + B\sin(\lambda r)],$$

where A and B are constants. If we apply (4.5.2) to (4.5.4), we immediately get $A = 0$, and so

$$v(r, t) = Be^{-\lambda^2 kt} \sin(\lambda r),$$

which says

(4.5.5) $$u = \frac{v}{r} = \frac{Be^{-\lambda^2 kt} \sin(\lambda r)}{r}.$$

So far, this should all be old hat to you.

Now we apply the boundary condition (4.5.3) at the insulated surface. This quickly reduces (4.5.5) to (you should confirm this claim)

$$\lambda R \cos(\lambda R) - \sin(\lambda R) = 0.$$

That is, the constant λ is *not* arbitrary, but instead is the solution to the transcendental equation

$$(4.5.6) \qquad\qquad \tan(\lambda R) = \lambda R,$$

solvable either by numerical means or, as Fourier (who, alas, lacked access to a modern electronic computer) suggests[13] in *Analytical Theory*, by an iterative (very laborious) graphical approach. In any case, there are an infinite number of solutions, as the line $y = \lambda R$ intersects branch after branch of the periodic function $y = \tan(\lambda R)$, as shown in Figure 4.5.1. Each such intersection gives a positive value for λ (we ignore the obvious $\lambda = 0$ solution, which (4.5.5) shows gives the trivial solution $u = 0$), which we'll label as λ_n, $n = 1, 2, 3, \ldots$. Since each value of B can be different for each of these values of λ, we then have for the general solution, with reference to (4.5.5):

$$(4.5.7) \qquad\qquad u(r, t) = \sum_{n=1}^{\infty} B_n \frac{e^{-\lambda_n^2 k t} \sin(\lambda_n r)}{r}.$$

The final question to answer is: what are the values of B_n? If we are given the initial temperature distribution in the sphere, say

$$(4.5.8) \qquad\qquad u(r, 0) = f(r),$$

then (4.5.7) becomes

$$(4.5.9) \qquad\qquad rf(r) = \sum_{n=1}^{\infty} B_n \sin(\lambda_n r).$$

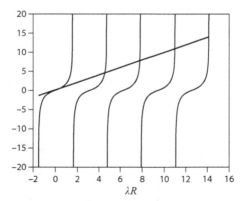

FIGURE 4.5.1. Solving for the λ_n.

Notice, carefully, that the right-hand side of (4.5.9) is *not* a Fourier series. That's because the λ_n for $n > 1$ are not integer multiples of λ_1, as would be the case in a true Fourier series.[14]

Nonetheless, there is an elegant way to compute the B_n. If we multiply (4.5.9) by $\sin(\lambda_m r)$, where m is some (any) particular value selected from the positive integers, and then integrate from zero to R, we have

$$(4.5.10) \quad \int_0^R rf(r)\sin(\lambda_m r)\,dr = \sum_{n=1}^{\infty} B_n \int_0^R \sin(\lambda_n r)\sin(\lambda_m r)\,dr.$$

Now, whatever value m has, as n varies in the sum on the right from 1 to infinity, it will, at some point, equal m. That particular integral in the sum is

$$\int_0^R \sin^2(\lambda_m r)\,dr,$$

which clearly has a well-defined, nonzero value for the known values of R and λ_m. What makes this useful is that all the other integrals in the sum, for $n \neq m$, *vanish*. If that's true, then the coefficient B_m is given by

$$(4.5.11) \quad B_m = \frac{\int_0^R rf(r)\sin(\lambda_m r)\,dr}{\int_0^R \sin^2(\lambda_m r)\,dr}.$$

Okay, *how* do we know all those other integrals vanish? Here's how to show that. Since trigonometry tells us

$$\sin(\lambda_n r)\sin(\lambda_m r) = \frac{1}{2}[\cos\{(\lambda_n - \lambda_m)r\} - \cos\{(\lambda_n + \lambda_m)r\}],$$

then the rightmost integral in (4.5.10) becomes (remember, we are now assuming $n \neq m$ and so $\lambda_n \neq \lambda_m$)

$$\int_0^R \sin(\lambda_n r)\sin(\lambda_m r)\,dr = \frac{1}{2}\left[\int_0^R \cos\{(\lambda_n - \lambda_m)r\}dr - \int_0^R \cos\{(\lambda_n + \lambda_m)r\}dr\right]$$

$$= \frac{1}{2}\left[\left.\frac{\sin\{(\lambda_n - \lambda_m)r\}}{\lambda_n - \lambda_m}\right|_0^R - \left.\frac{\sin\{(\lambda_n + \lambda_m)r\}}{\lambda_n + \lambda_m}\right|_0^R\right]$$

$$= \frac{1}{2}\left[\frac{\sin\{(\lambda_n - \lambda_m)R\}}{\lambda_n - \lambda_m} - \frac{\sin\{(\lambda_n + \lambda_m)R\}}{\lambda_n + \lambda_m}\right].$$

Next, remembering the identity $\sin(A \pm B) = \sin(A)\cos(B) \pm \cos(A)\sin(B)$, we have

$$\int_0^R \sin(\lambda_n r)\sin(\lambda_m r)\,dr = \frac{1}{2}\left[\frac{\sin(\lambda_n R)\cos(\lambda_m R) - \cos(\lambda_n R)\sin(\lambda_m R)}{\lambda_n - \lambda_m} - \frac{\sin(\lambda_n R)\cos(\lambda_m R) + \cos(\lambda_n R)\sin(\lambda_m R)}{\lambda_n + \lambda_m}\right].$$

From (4.5.6), we have

$$\tan(\lambda R) = \frac{\sin(\lambda R)}{\cos(\lambda R)} = \lambda R,$$

and so we can replace any $\sin(\lambda R)$ factor with $\lambda R\cos(\lambda R)$. Thus,

$$\int_0^R \sin(\lambda_n r)\sin(\lambda_m r)\,dr$$

$$= \frac{1}{2}\left[\frac{\lambda_n R\cos(\lambda_n R)\cos(\lambda_m R) - \lambda_m R\cos(\lambda_m R)\cos(\lambda_n R)}{\lambda_n - \lambda_m} - \frac{\lambda_n R\cos(\lambda_n R)\cos(\lambda_m R) + \lambda_m R\cos(\lambda_m R)\cos(\lambda_n R)}{\lambda_n + \lambda_m}\right]$$

$$= \frac{R}{2}\left[\frac{(\lambda_n - \lambda_m)\cos(\lambda_n R)\cos(\lambda_m R)}{\lambda_n - \lambda_m} - \frac{(\lambda_n + \lambda_m)\cos(\lambda_n R)\cos(\lambda_m R)}{\lambda_n + \lambda_m}\right] = 0$$

after doing the obvious cancellations.[15] This has all been easy, high school algebra, but the end result is not, I believe, at all obvious. With the B_m now determined in (4.5.11), (4.5.7) solves the insulated sphere problem.

Well, it's been a *very* mathematical adventure up to now but, at last, you are ready for the Atlantic cable. So, off we go to the bottom of the sea, where despite conditions being considerably cooler than they once were in the molten Earth of Chapter 3 (or the heated ring of the previous section)—and, in any case, we will be discussing electrical energy and not heat energy—we'll happily find that the heat equation still applies.

CHAPTER 5

William Thomson and the Infinitely Long Telegraph Cable Equation

5.1 The Origin of the Atlantic Cable Project

I think it highly likely that anyone who would read a book like this one has also read many of the classic Victorian adventure novels at some time during their youth. I'm thinking, for example, of Jules Verne's *From the Earth to the Moon* (1865), Verne's *Around the World in Eighty Days* (1873), and H. G. Wells's *The Time Machine* (1895). Verne and Wells were great literary competitors, with distinctly different approaches[1] to the "scientific romance" (as their genre was known, before the term *science fiction* was coined), but these three novels (and others of the same ilk) shared one common feature: all start with a seemingly outrageous project being vigorously debated among a group of clearly intelligent, upper-class men of means expressing varying degrees of skepticism about the possibility of the project.

In *From the Earth to the Moon*, the debate, at the fictional Baltimore Gun Club, swirls around the possibility of using a giant cannon to blast a manned capsule to the Moon. In *Around the World in Eighty Days* the debate at London's Reform Club (an actual club in existence today) is (as you might guess from the title) about being able to complete a journey around the globe in 80 days. And Wells's fantastic novel opens with a dinner party at the Time Traveler's home, after which he demonstrates a minature working model of his time machine to his astonished (and still skeptical) guests.

Wild and crazy, but wonderful, too, were these novels for their Victorian readers (the first two projects have, of course, long been accomplished by real adventurers, while the third still waits for one or more crucial breakthroughs in scientific understanding). In 1854, however, a real-life enterprise was similarly born that Verne, himself, might have hesitated in giving a fictional treatment: the laying of a 2,000-mile-long cable nearly 3 miles beneath the surface of the Atlantic ocean, a cable to allow "almost instantaneous" electrical communication between the telegraph landlines of Ireland in the Old World and those of Newfoundland in the New World. The construction of this cable, weighing 10 million pounds, was rightfully considered to be one of the greatest achievements of 19th-century engineering and science, and understanding how it worked depended on being able to solve, of all things, Fourier's heat equation.

The need (if not the possibility of) for such an enormous cable[2] was quite clear by the mid-19th century. In our massively connected world of the 21st century (a world in which you can flip on your electronic tablet and watch a live broadcast originating in downtown Berlin as you drink a cup of coffee while sitting on your living room sofa in Los Angeles[3]), it can be difficult to appreciate just how *disconnected* the world was two centuries ago. When King George II died in 1760, for example, it took 6 weeks (!) for the news to reach colonial America. And more than a century later, when Abraham Lincoln was assassinated in 1865, it still took nearly 2 weeks for the news to travel in the other direction to London. Contrast that to 2001, when people world-wide witnessed as it happened the deaths of thousands as the World Trade Center Towers in New York City were destroyed by terrorists.

The originators of the Atlantic cable project were capitalists who were certainly motivated by the possibility of earning impressive returns on their financial investments.[4] But there can be no doubt, too, that even those shrewd businessmen were caught up in the sheer thrill of attempting what more ordinary folk would call a foolish, indeed an *impossible*, task. The major concern for them was the obvious one, of course, that of overcoming the immense water barrier of the mighty Atlantic Ocean. The mystery of what lay thousands of feet beneath its heaving, stormy surface was quite enough to give even the most adventurous of individuals reason enough to hesitate. After all, as so many argued, just think of the incredible pressure the cable would experience at a depth of 15,000 feet!

Water pressure increases (roughly) at a rate of a little less than half-a-pound per square inch for each foot of depth, and at 15,000 feet, the pressure on the cable would be more than 6,600 pounds per square inch.

(Even modern nuclear submarines can't withstand such a fearsome pressure: when the *USS Thresher* sank in 1963, it imploded at a depth of about 2,400 feet.) And so one amusing Victorian concern was that such an "uncommon compression" would result in a robust electrical signal going into the sending end of the cable but emerging at the receiving end only after having been reduced to a "mouse-like squeaking." A concern about deep-water pressure found its way into the popular literature of the day, with Jules Verne giving a nice mathematical exposition on the issue in his famous 1869 novel *20,000 Leagues under the Sea.*

There were other concerns. The cable, for example, was often likened to being a mere thread to be tossed into the vastness of the ocean: but how could such an ephemeral object actually sink to the bottom of the sea? Thinking that the density of water increases with depth because of the increasing pressure (which doesn't happen, because water is incompressible), and then invoking Archimedes's principle, it was thought that sinking objects would stop descending once they had displaced their own weight. So, wouldn't the cable sink only a little and then just hang in the water, open to all the multitude of forces attempting to break it (whales ascending from the depths, icebergs, ship anchors, sharks, and other hideous sea monsters, all with mouths filled with viciously sharp teeth, being just a few of the threats commonly conjured up by those with hyperactive imaginations)?

Despite these and other widespread worries (known to the men of science of the times to be mostly—but maybe not completely—in error), in January 1854, the idea of creating what would be the start of a worldwide Victorian-era "internet" reached the ears of a young, wealthy, highly connected American in New York City, one Cyrus West Field (1819–1892). He quickly convened a meeting of other potential investors (just like in the opening of a Jules Verne novel!) in March, and so was born the Atlantic Cable Company, with a pile of pre–Civil War cash in the amount of $1,500,000 behind it.

And after just 12 years of high adventure in both the worlds of engineering and finance (another 4 million dollars was raised and spent), the Atlantic cable finally went into regular, long-term operation.[5] Earlier attempts, starting in 1857, had ultimately failed, but one particularly gushing oration by a prominent statesman and educator, just after the first try (which, briefly, actually worked), illustrates how magnificent an achievement the cable was thought to be:

Does it seem all but incredible to you that intelligence should travel for two thousand miles, along those slender copper wires, far down in the all but fathomless Atlantic; never before penetrated . . . save when some floundering vessel has plunged with hapless company to the eternal silence and darkness of the abyss? Does it seem . . . but a miracle . . . that the thoughts of living men . . . far down among the uncouth monsters that wallow in the nether seas, along the wrecked paved floor, through the oozy dungeons of the rayless deep . . . should go flashing along the slimy decks of old sunken galleons, which have been rotting for ages; that messages of friendship and love from warm living bosoms, should burn over the cold, green bones of men and women whose hearts, once as warm as ours, burst as the eternal gulfs closed and roared over them centuries ago?[6]

Pretty gruesome stuff, all right, but certainly heartfelt!

5.2 Some Electrical Physics for Mathematicians

So, the cable made some reputations,[7] as well as a lot of money. Our goal, from this point on, is to understand how the cable *worked*. Just to be sure I don't fall into the fault of using a lot of electrical engineering jargon, to the irritation of those readers who did not spend their high school years inhaling the intoxicating fumes of molten solder and bubbling rosin flux (and who also missed out on experiencing the excruciating pain, to say nothing of the embarrassment, of accidentally grabbing the wrong end of a hot soldering iron), I'll summarize here what the rest of this chapter will assume as background. If this is all old hat to you, however, just skip ahead to the next section.

There are three fundamental components commonly used in electrical/electronic circuitry: resistors, capacitors, and inductors. All of these components are passive. That is, they do not generate electrical energy, but rather either *dissipate* energy as heat (resistors), or temporarily *store* energy in an electric field (capacitors) or in a magnetic field (inductors). All three components have two terminals, as shown in Figure 5.2.1.

There are, of course, other more complex, multi-terminal components used in electrical/electronic circuits (such as transistors), as well as such things as constant voltage, and constant current, sources,[8] but for our purposes, these three will be all we'll need. (Actually, as you'll see, only

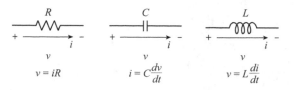

FIGURE 5.2.1. The three standard, passive, two-terminal components: the resistor, capacitor, and inductor.

the first two will appear in our discussion of the Atlantic cable.) We can formally define each of the three passive, two-terminal components by the relationship that connects the current i through them to the voltage drop v across them, as is done in Figure 5.2.1. If we denote the values of resistors, capacitors, and inductors, respectively, by R (ohms), C (farads), and L (henrys), and if v and i have the unit of volts and amperes (named after the Italian scientist Alessandro Volta (1745–1827) and the French mathematical physicist André Marie Ampere (1775–1836), respectively), and if time t is in units of seconds, then the mathematical definitions of the components are as shown in Figure 5.2.1.

Of course, $v = iR$ is the famous *Ohm's law*, named after the German Georg Ohm (1787–1854). The other two relationships haven't been given commonly used names, but the units of capacitance and inductance are named, respectively, after the English experimenter Michael Faraday (1791–1867) and the American physicist Joseph Henry (1799–1878). As a general guide, 1 ohm is a small resistance, 1 farad is very large capacitance, and 1 henry is a fairly large inductance. The possible ranges on voltages and currents in actual circuitry is enormous, ranging from micro-volts/micro-amps to mega-volts/mega-amps. For the Atlantic cable, the voltages at the sending end were measured in a few dozens of volts, while the currents at the receiving end were in the milliamp range.

The mathematical current-voltage laws of the resistor and the capacitor are sufficient in themselves for what we'll do in this chapter (that is, we don't really need to delve more deeply into how they work, but I will in fact say just a bit more about each in just a moment). In contrast, we can temporarily forget about inductors, since the Atlantic cable was purposely modeled to be *non*-inductive. (But in Chapter 6, I will bring induction back into our discussion.)

When William Thomson performed his theoretical analysis of the Atlantic cable, he used two "laws," dating from 1845, named after the German physicist Gustav Robert Kirchhoff (1824–1887). These two laws,

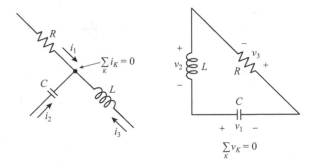

FIGURE 5.2.2. Kirchhoff's two circuit laws.

illustrated in Figure 5.2.2, are in fact actually the fundamental physical laws of the conservation of energy and the conservation of electric charge. They are easy to state.

Kirchhoff's voltage law: The sum of the *voltage (or electric potential) drops* around any closed path (loop) in a circuit is zero. Voltage is defined to be energy per unit charge; the voltage *drop* is the energy expended in transporting a unit charge through the electric field that exists inside the component. The law, then, says that the net energy change for a unit charge transported around a closed path is zero. If it were not zero, then we could repeatedly transport charge around the closed path in the direction in which the net energy change is positive and so become rich selling the energy gained to the local power company. Conservation of energy (and a denial of the possibility of constructing a perpetual motion machine), however, says we can't do that. (Since the sum of the drops is zero, then one can also set the sum of the voltage *rises* around any closed loop to zero.)

Kirchhoff's current law: The sum of the currents into any point in a circuit is zero. This says that if we construct a tiny, closed surface around any point in a circuit, then the charge enclosed by that surface remains constant. That is, whatever charge is transported into the enclosed volume by one current is transported out of the volume by other currents; current *is* the motion of electric charge. Mathematically, the current i at any point in a circuit is defined to be the rate at which charge is moving through that point, that is, $i = dQ/dt$. Q is measured in coulombs—named after the French physicist Charles Coulomb (1736–1806)—where the charge on an electron is $e = -1.6 \times 10^{-19}$ coulombs.[9] One ampere is 1 coulomb per second.

The two laws allow us to make a couple of highly useful, general statements concerning resistors:

1. Resistors add when connected in series because they are each carrying the same current (then apply Kirchhoff's voltage law). That is, the combined resistance is $R = R_1 + R_2 + \cdots$.
2. Resistors combine when connected in parallel as $\frac{1}{R} = \frac{1}{R_1} + \frac{1}{R_2} + \cdots$, because they each have the same voltage drop across their terminals (then apply Kirchhoff's current law).

These two statements might seem to be quite limited, as electrical circuits are not made only from resistors, but at the end of this section, I'll show you how these two statements still do have applicability to circuits containing inductors and capacitors in addition to resistors.

A common, popular misconception about electricity in wires is that "things move at the speed of light." While it is true that *information* travels at speeds measured in miles per second (but still less than the speed of light, which is equal to 3×10^8 meters per second), physical electrons in wires hardly move at all! We'll determine the information speed in the Atlantic cable later in this chapter, but we can calculate the electron speeds right now. Suppose the density of electrons available in a wire (the so-called *conduction electrons* that come from the weakly bound outer-orbit valence electrons of the wire's atoms) is n electrons per cubic meter. Then the conduction charge *density* is $\rho = ne$. If the wire has a uniform cross-sectional area of A square meters, and if s is the speed of the conduction electrons (in meters per second) when the current is i amperes, then we see that, dimensionally, we must have $i = \rho s A$. That is,

$$s = \frac{i}{\rho A}.$$

For example, suppose we have some #30 copper wire, which has a circular cross-sectional diameter of 0.255 mm, and $n = 8.43 \times 10^{28}$. We calculate the electron speed for a current of 5 mA (typical of a received current on the Atlantic cable) as follows:

$$s = \frac{0.005 \text{ coulombs/s}}{8.43 \times 10^{28} \frac{\text{electrons}}{\text{m}^3} \times 1.6 \times 10^{-19} \frac{\text{coulombs}}{\text{electron}} \times \frac{\pi}{4} (0.255 \times 10^{-3})^2 \text{m}^2}$$

$$= 7.26 \times 10^{-6} \frac{\text{m}}{\text{s}}.$$

That's just one inch per hour. That's pretty slow! Indeed, the electron conduction speed, even for a reasonably large current like 5 mA, is often called the *drift* speed.

After reading these last several paragraphs, you might have been lulled into thinking "Well, those electrical fellows have pretty much explained electricity." *Nothing could be further from the truth.* Here, to explain what I mean by that, is a wonderful little story I liked to tell my electrical engineering students when I was still teaching, a story that reveals the real situation. It may be apocryphal, but even if it is, it honestly reflects the present state of our (lack of) understanding:[10]

> In Cambridge, they tell the following story about Maxwell:[11] Maxwell was lecturing and, seeing a student dozing off, awakened him, asking "Young man, what is electricity?" "I'm terribly sorry, sir," the student replied, "I knew the answer but I have forgotten it." Maxwell's response to the class was, "Gentlemen, you have just witnessed the greatest tragedy in the history of science. The one person who knew what electricity is has forgotten it."

Surely, you object, a scientist as great as Maxwell must have had *some* idea of what electricity is, and you'd be right. He (at one time) thought it to be like an incompressible fluid, a view that lasted quite a while in Victorian times. Indeed, that was the view held by Thomson when he analyzed the Atlantic cable. That, however, is just a description of how electricity was thought to behave; it doesn't say anything about what it *is*, and that remains a mystery to this day. A similar situation existed in Victorian times concerning gravity. Everybody since Newton knew how gravity *behaved*, but it wasn't until Einstein's general relativity, and the ideas of space-time and its *curvature* were developed, that the understanding of gravity moved beyond that of merely describing it (the inverse-square law). A similar revelation about electricity lies in the future.[12]

Now, just a few more elaborative words about electrical matters in general, and resistors and capacitors in particular, the "components" out of which the Atlantic cable was constructed. I've put *components* in scare quotes, because the cable was not actually made from *discrete* (or *lumped*) resistors and capacitors, but rather from what electrical engineers call *distributed* values (more on that in the next section).

Electrons, with negative charge, travel in the direction opposite to the direction of what is called the *conventional current*, which flows from a point with "high" voltage to a point with "low" voltage, just as heat energy flows from high-temperature to low-temperature regions. This is because the conventional current is imagined to be the flow of *positive* charge carriers. That is, the conventional current flows in the direction of the voltage drop, while the electron current does the opposite (and you can blame Ben Franklin and his famous kite in a lightning storm experiment for this odd bit of common confusion).

As another example of the sometimes awkward state of affairs in our understanding of electricity, I've already mentioned something called the *electric field* several times in this section, and have depended on you not to be (too) bothered by it. But what is an electric field? It is a hugely convenient *mathematical* artifice (measured in units of volts per meter, the voltage drop from point **a** to point **b** in a circuit is the integral of the field from **a** to **b**), but it is a great *physical* mystery. What's different about the empty space in which a field exists, compared to the state of that same empty space when the field doesn't exist? Nobody knows. As I'll show you at the end of this section, the electric field inside a capacitor is thought to be able to store energy, and not to dissipate it as heat energy as does a resistor. *Where* and *how* is that energy stored? Nobody knows. But there can be no doubt as to the physical reality of that energy, as anybody who has ever carelessly stuck his or her hand into an unplugged electrical circuit containing a capacitor charged to several hundred volts will testify!

Electrons, having mass, need energy to get them moving, and that energy comes from the electric field that exists throughout the conductors and components of an energized circuit. The field exerts an accelerating force on the conduction electrons, but as they flow, the electrons continuously collide with the immobile atoms of the material in which they (the electrons) travel. With each collision, they give up the energy gained since their last collision, and so this acceleration/collision/loss of energy process repeats, over and over. This continual loss of energy heats the bulk material of the immobile atoms, and so the temperature of resistors carrying current rises.

Now, recalling the definition of the voltage drop between two points as the energy expended in transporting a unit positive charge through the electric field between those two points, it then follows that if we transport a charge ΔQ through a voltage drop of v then the total energy (measured in joules, after the English physicist James Joule (1818–1889)) is $v\Delta Q$. If the time required to transport ΔQ is ΔT, then the energy per unit time or *power p* (measured in watts, after the Scottish engineer James Watt (1736–1819)), is

$$(5.2.1) \qquad p = \frac{v\Delta Q}{\Delta T} = v\frac{dQ}{dt} = vi,$$

where of course i is the current. One volt-ampere is one watt, which is 1 joule per second. Writing (5.2.1) explicitly as instantaneous functions of time,

$$(5.2.2) \qquad p(t) = v(t)i(t),$$

and so, in a resistor where $v(t) = i(t)R$, we have

$$i(t) = \frac{v(t)}{R},$$

which says the power in a resistor (the rate at which the resistor dissipates electrical energy as heat energy) is

$$(5.2.3) \qquad p(t) = v(t)\frac{v(t)}{R} = \frac{v^2(t)}{R}.$$

Equivalently,

$$(5.2.4) \qquad p(t) = [i(t)R]i(t) = i^2(t)R.$$

In either case, if $R = 1$ ohm, we see from (5.2.3) and (5.2.4) that $p(t)$ is the *square* of a time function, and so the total energy dissipated by the resistor during any time interval is the integral with respect to time of the square of a time function. You'll recall that back in Chapter 2—

right after (2.1.11)—that's exactly the definition for the energy of a time function we used, and now you see the motivation for doing that. Our *mathematical* definition was motivated by *hot resistors*—and isn't that perfectly appropriate in a book on Fourier's heat equation?

Now, let's turn our attention to capacitors. If a circuit is not complete with a solid conductive path, but instead contains an air gap formed by a pair of parallel plates, a current $i(t)$ can still flow. That current will deposit charge on each plate: one plate will have a negative charge $-Q$ (consisting of electrons), and the other plate will have a positive charge $+Q$ (consisting of ionized atoms missing some of their valence electrons, indeed, the same number of missing electrons as the number of electrons that are on the other plate). These two charged plates are imagined to create an electric field in the space between the plates, and so there will be a voltage drop v from the positive plate to the negative plate. It is found experimentally that v and Q are proportional: the proportionality constant is called the *capacitance* C. That is,

$$Cv = Q,$$

or differentiating,

$$\frac{d}{dt}(Cv) = \frac{dQ}{dt} = C\frac{dv}{dt},$$

or because $i = \frac{dQ}{dt}$, we have

$$(5.2.5) \qquad i(t) = C\frac{dv}{dt},$$

as shown in Figure 5.2.1.

Recall that at the start of this section, I said that, unlike resistors, capacitors don't dissipate energy. Here's why that's so. The power in a capacitor is, from (5.2.2) and (5.2.5),

$$p(t) = v(t)i(t) = vC\frac{dv}{dt} = \frac{1}{2}C\frac{d(v^2)}{dt},$$

and so the energy delivered to the capacitor over the time interval 0 to T is given by

$$\int_0^T p(t)\,dt = \frac{1}{2}C\int_0^T \frac{d(v^2)}{dt}\,dt = \frac{1}{2}C\int_0^T d(v^2).$$

The limits on the last integral are units of time, while the variable of integration is v^2. What we need to do is change the limits on the integral to match v^2, and so the total energy is

$$\frac{1}{2}C\int_{v^2(0)}^{v^2(T)} d(v^2) = \frac{1}{2}C[v^2(T) - v^2(0)].$$

Notice that this energy is zero if $v(0) = v(T)$. That is, all the energy *delivered* to the capacitor is *returned* to the circuit if the voltage drop across the capacitor is the same at $t = T$ as it was at $t = 0$. Thus, the delivered energy was not dissipated, but must have been (somehow) stored. As I said earlier, just how and where this storage is accomplished remains a mystery.

For a resistor R, the ratio of the voltage drop across the resistor to the current in it is *always* a constant (specifically, the ratio is R). Because of the presence of a differentiation operator in their voltage-current laws, however, this is not true, in general, for inductors and capacitors. If we write $p = \frac{d}{dt}$ as the so-called *differentiation operator*, then we can write

$v = iR$ (resistor R)
$i = Cpv$ (capacitor C)
$v = Lpi$ (inductor L).

If we define the ratio $Z = \frac{v}{i}$ as the "resistance" (the technical name is *impedance*) of a circuit element, then

$Z = R$ (resistor)
$Z = \frac{1}{Cp}$ (capacitor)
$Z = Lp$ (inductor).

The merit in doing this is that it allows inductors and capacitors in complicated circuits to be treated (mathematically) just as if they were resistors. So, the impedance of a resistor R in series with an inductor L is $Z = R + Lp$, and the impedance of a capacitor C in parallel with an inductor L is

$$\frac{1}{Z} = \frac{1}{\dfrac{1}{Cp}} + \frac{1}{Lp} = Cp + \frac{1}{Lp} = \frac{LCp^2 + 1}{Lp},$$

which gives

$$Z = \frac{Lp}{LCp^2 + 1} = \left(\frac{1}{C}\right) \frac{p}{p^2 + \dfrac{1}{LC}}.$$

Now, to show you an elegant application of this electrical physics, combined with an equally elegant mathematical argument, consider the infinitely long circuit of Figure 5.2.3, constructed from an infinite number of identical subcircuits connected in series. The interior of each subcircuit is shown in Figure 5.2.4.

For us, now, Figure 5.2.4 is just a random example, but in fact you'll see in Chapter 6 that the subcircuit is the model for a *short* (Δx) length of telegraph cable, where R, L, and C are the cable resistance, inductance, and capacitance *per unit length*, and G represents a leakage resistance per unit length between the two conductors of the cable (actually, G is the reciprocal of the leakage resistance per unit length, and $G\Delta x$ is called a *conductance*). Thus, a long cable, as in Figure 5.2.3, is the series connection of a great many (*infinitely* many, as $\Delta x \to 0$) of the subcircuits of Figure 5.2.4.

Now, suppose we apply an *arbitrary* voltage $v(t)$ to the terminals a, b at the far left of Figure 5.2.3. This applied voltage can have any time behavior we wish, and we imagine the resulting input current is $i(t)$. This probably all seems to be so general that the following claim may be hard to believe: it is possible to adjust the values of R, L, C, and G so that the ratio $\frac{v(t)}{i(t)}$ is a constant, *independent of time*. This result had tremendous

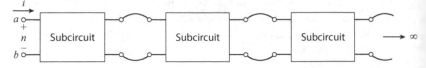

FIGURE 5.2.3. An infinitely long circuit.

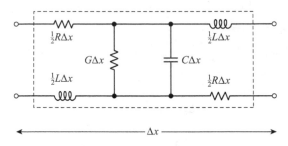

FIGURE 5.2.4. The interior of a subcircuit.

significance for the development of the trans-Atlantic cables that came after the original Atlantic cable, and we'll return to this in Chapter 6. For now, let's simply establish the claim.

To start, I'll make a borderline metaphysical argument that I think you'll easily accept as being plausible. If we write Z as the impedance at the input that we see at the terminals a, b of the subcircuit at the far left of Figure 5.2.3, then, since the long circuit of Figure 5.2.3 is *infinitely* long, removing a tiny portion of the circuit at the input should have no effect on the impedance we see. That is, we should see the *same Z* at the input terminals of the second subcircuit, and so we can replace the infinite cable from the second subcircuit onward by the impedance Z, to arrive at Figure 5.2.5.

When going from Figure 5.2.4 to Figure 5.2.5, remember that $G\Delta x$ is a conductance and so has an impedance of $\frac{1}{G\Delta x}$, that the capacitance $C\Delta x$ has an impedance of $\frac{1}{C\Delta xp}$, and that the inductance $L\Delta x$ has an impedance of $L\Delta xp$. Let the three parallel paths at the right of Figure 5.2.5 have impedance Z', so that

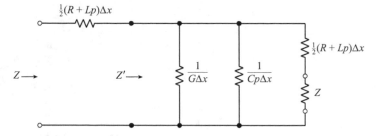

FIGURE 5.2.5. The electrical equivalent of Figure 5.2.3.

$$\frac{1}{Z'} = \frac{1}{\dfrac{1}{G\Delta x}} + \frac{1}{\dfrac{1}{C\Delta xp}} + \frac{1}{\dfrac{1}{2}(R+Lp)\Delta x + Z},$$

or

$$\frac{1}{Z'} = G\Delta x + Cp\Delta x + \frac{1}{\dfrac{1}{2}(R+Lp)\Delta x + Z}$$

$$= \frac{(G+Cp)\Delta x\left[\dfrac{1}{2}(R+Lp)\Delta x + Z\right] + 1}{\dfrac{1}{2}(R+Lp)\Delta x + Z},$$

or

$$Z' = \frac{\dfrac{1}{2}(R+Lp)\Delta x + Z}{(G+Cp)\dfrac{1}{2}(R+Lp)(\Delta x)^2 + Z(G+Cp)\Delta x + 1}.$$

Now, Δx is imagined to be very short and so, ignoring terms in $(\Delta x)^2$ in comparison to those in Δx, we can write (as Δx becomes arbitrarily small)

$$Z' = \frac{\frac{1}{2}(R+Lp)\Delta x + Z}{Z(G+Cp)\Delta x + 1}.$$

Since

$$Z = \frac{1}{2}(R+Lp)\Delta x + Z' = \frac{1}{2}(R+Lp)\Delta x + \frac{\frac{1}{2}(R+Lp)\Delta x + Z}{Z(G+Cp)\Delta x + 1},$$

we have

$$Z^2(G+Cp)\Delta x + Z = Z\tfrac{1}{2}(R+Lp)(G+Cp)(\Delta x)^2 + \tfrac{1}{2}(R+Lp)\Delta x$$
$$+ \tfrac{1}{2}(R+Lp)\Delta x + Z,$$

or again ignoring terms in $(\Delta x)^2$ in comparison to those in Δx,

$$Z^2(G+Cp)\Delta x = (R+Lp)\Delta x,$$

or

$$Z^2 = \frac{R+Lp}{G+Cp} = \frac{L\left(p+\dfrac{R}{L}\right)}{C\left(p+\dfrac{G}{C}\right)}.$$

That is,

$$Z = \sqrt{\frac{L}{C}}\sqrt{\frac{p+\dfrac{R}{L}}{p+\dfrac{G}{C}}} = \frac{v}{i}.$$

Now, before going any further, ask yourself just what such a Z *means*? Since $Z = \frac{v}{i}$ then what it means is

$$i(t) = \left\{ \sqrt{\frac{C}{L}} \sqrt{\frac{p + \dfrac{G}{C}}{p + \dfrac{R}{L}}} \right\} v(t),$$

where the expression on the right in the curly brackets is an operator (of a decidedly nontrivial nature) working on $v(t)$ to give $i(t)$. Constructed from our original $p = \frac{d}{dt}$ operator (which seems clear enough), the question of how to physically interpret this new, not-so-clear operator was a hot topic in 19th-century mathematics (try googling *fractional operators* and see what you get). We'll completely sidestep that question by eliminating the presence of p in the operator! To do that, simply notice that if

(5.2.6) $$\frac{R}{L} = \frac{G}{C},$$

then Z becomes a real constant (no p dependency). That is, the impedance of the infinite circuit of Figure 5.2.3 is the pure *resistance*,

$$Z = \sqrt{\frac{L}{C}},$$

when (5.2.6) is satisfied. This is pretty amazing, given that our circuit contains inductors and capacitors, as well as resistors. For now, this is simply an interesting mathematical result, but I'll remind you of (5.2.6) when we get to Chapter 6, because it had enormous practical implications in telegraph cable technology as well.

Okay, *now, at last*, we come to the Atlantic cable.

5.3 Derivation of the Heat Equation as the Atlantic Cable Equation

This section will be mostly mathematical, but before we start, let me tell you about one additional, early objection to the entire *idea* of an Atlantic cable. The concerns I mentioned in the opening section of this chapter were at least of a physical nature: this new concern was an *emotional* one, and that's the reason I've put it here instead of earlier. Proceeding with the cable despite this particular concern was done—to use a phrase attributed to the American theoretical physicist Robert Oppenheimer (1904–1967)—because the Atlantic cable was just too "technically sweet" not to make it.[13]

The objection was that the primary use of the cable would be to send trivial messages! Here, for example, is what the *New York Times* had to say (in August 1858) about a trans-Atlantic cable: "Superficial, sudden, unsifted, too fast for the truth, must be all telegraphic intelligence. Does it not render the popular mind too fast for the truth? Ten days bring us the mails from Europe. What need is there for the scraps of news in ten minutes?" A few years later, the paper was grumbling again about how telegraphy, in general, was simply too fast: telegraphy made, it was asserted, people "mourn for the good old times when mails came by steamer twice a month." With no little irony, one reads similar objections today about modern social media platforms, like Twitter and Internet blogs! Well, no matter what the nature of the Atlantic cable messages might be, William Thomson for one found the theoretical challenges presented by the cable just too technically sweet to ignore. Here's what he did.

Figure 5.3.1 shows the geometry of a two-conductor cable: the figure displays an arbitrary, very short (Δx) section of the cable, which has what are called *distributed parameters*:

$$R = \text{resistance per unit length (due to the resistance}$$
$$\text{of the two conductors)}$$

and

$$C = \text{capacitance per unit length (due to the electric}$$
$$\text{field between the two conductors).}$$

The cable is imagined as consisting of a cascade-in-series of a very large number of these sections (eventually we'll let $\Delta x \to 0$).

The cable is depicted as having two conductors to form a complete circuit, but this is strictly for ease of discussion. The early cables actually used the ocean itself as the return path to complete the circuit. Typical

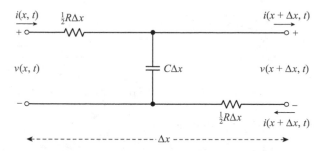

FIGURE 5.3.1. An arbitrary, very short section of the Atlantic cable.

values for the distributed resistance and capacitance on the Victorian cables were 3 ohms/nautical mile and 0.5 microfarads/nautical mile, respectively. These numbers are consistent with the statement by the English mathematical physicist John William Strutt (1842–1919), better known in the world of physics as Lord Rayleigh (a friend of Thomson, and as the 1904 Nobel physics prize winner, every bit his scientific equal), that the RC product on the Atlantic cables was on the order of 5×10^{-17} in cgs units (s/cm^2).[14]

The circuit in Figure 5.3.1 is drawn to depict the cable as being electrically symmetrical—that is, to look the same from left-to-right and right-to-left—and that is accomplished by placing half of the cable resistance in the top conductor on the left (into which flows the current $i(x, t)$), and the other half of the cable resistance in the bottom (return) conductor on the right (into which flows the "return" current $i(x + \Delta x, t)$). The cable capacitance is modeled by placing it in the middle of the circuit, which preserves symmetry. The circuit, of course, has no distributed inductance *by assumption*. And finally, the Atlantic cable was also assumed to have no wire-to-wire *leakage*, which would be modeled as a resistor in parallel with the C. Such a resistance was actually present in a real cable but, as it was generally in the millions of ohms there was little error in taking it as infinite. (The inductance and leakage assumptions will be removed in Chapter 6.)

Starting at the upper-left point, where the current $i(x, t)$ enters the upper-left $\frac{1}{2} R \Delta x$ resistance, and summing voltage drops around the outer closed loop of the circuit, Kirchhoff's voltage law says

$$i(x,t)\tfrac{1}{2}R\Delta x + v(x+\Delta x,t) + i(x+\Delta x,t)\tfrac{1}{2}R\Delta x - v(x,t) = 0.$$

With a little rearranging,

$$-[v(x+\Delta x,t) - v(x,t)] = \tfrac{1}{2}R\Delta x[i(x,t) + i(x+\Delta x,t)].$$

Dividing through by Δx, we have

$$-\frac{v(x+\Delta x,t) - v(x,t)}{\Delta x} = \frac{1}{2}R[i(x,t) + i(x+\Delta x,t)].$$

If we now let $\Delta x \to 0$, we have $i(x+\Delta x, t) \to i(x, t) = i$, and so

(5.3.1)
$$-\frac{\partial v}{\partial x} = iR,$$

where a partial derivative is used because we have two independent variables.

Once we have (5.3.1) down on paper, it is clear (perhaps) that we should have been able to skip the math and to have argued directly that the *loss* of voltage (the physical meaning of the minus sign) along the cable per unit length is equal to the ohmic voltage drop per unit length. With that physical insight as inspiration, we can immediately write a similar equation for the loss of current along the cable per unit length as the current needed to charge the cable's capacitance between the two conductors. That is,

(5.3.2)
$$-\frac{\partial i}{\partial x} = C\frac{\partial v}{\partial t}.$$

If we differentiate (partially) (5.3.1) with respect to x, we have

$$-\frac{\partial^2 v}{\partial x^2} = R\frac{\partial i}{\partial x}.$$

Substituting (5.3.2) for $\frac{\partial i}{\partial x}$ into this, we get

(5.3.3)
$$\frac{1}{RC}\frac{\partial^2 v}{\partial x^2} = \frac{\partial v}{\partial t},$$

which is Fourier's *heat equation* (with $k = \frac{1}{RC}$)! So, all of Fourier's results for heat flow in a one-dimensional, semi-infinite rod, with its lateral surface insulated, immediately carry over for the behavior of *electricity* in a semi-infinite, induction-free telegraph cable, and this was William Thomson's great insight in 1855.[15]

Thomson's original derivation of (5.3.3) differs in the details from what I've shown you here, but not in substance. Reading Thomson in the original is still worthwhile today, as it offers a vivid illustration of how an ingenious mind works (in his paper, Thomson gives all due credit to Fourier's *Analytical Theory*). In addition, Thomson's 1855 paper is of particular interest to historians of science as it also contains excerpts from letters to Thomson, written by his close personal friend, the great English mathematical physicist George Gabriel Stokes (1819–1903), who was a professor at Cambridge. Dated late 1854, they show that Stokes, too, was greatly intrigued by the "technically sweet" theoretical challenges raised by the possibility of an Atlantic cable.

The impression you may have gotten from what you've just read is that all the technically sweet problems were electrical in nature, but that's not so. There were many other potential difficulties of a nonelectrical sort, too, and I'll mention just two to end this section. The *mechanical engineering* difficulties inherent in the Atlantic cable were daunting ones. Imagine, for example, the large, potentially destructive forces acting on the cable as it was passed off the deck of a rolling, pitching, yawing ship into the sea. There was the always-present danger of a cable break, with one end of a snapped cable (the successful 1866 cable had a breaking strength of 16,500 pounds!) murderously whip-slashing across the deck to cut a man in half before plunging into the depths. The on-deck machinery used to pay out the cable was called *the brake*, and as an indication of how the deck crewmen worried about cable snaps, here's what they'd sing as they worked (to the tune of "Pop Goes the Weasel"):

Pay it out, oh! Pay it out,
    As long as you are able;
For if you put the darned brakes on,
    Pop goes the cable!

When a break did occur, W. H. Russell (note 5) described the crew reaction as follows:

"[T]he Cable parted . . . and with one bound leaped over intervening space and flashed into the sea. The shock of the instant was as sharp as the snapping of the Cable itself. No words could describe the bitterness of the disappointment. The Cable was gone! Gone forever down in that fearful depth! It was enough to move one to tears." There then followed the infinitely frustrating task of grabbling for the lost end of the cable thousands of feet below, hauling it back up to the surface, and then splicing the two ends back together.

Another concern, which straddled the mechanical/electrical divide, was in the construction of the cable itself. The cable was going into a remote, hostile environment (who knew the details of what lay 10,000 feet and more at the bottom of the unexplored ocean?), where it was supposed to continue working for years, without human intervention. It had to be made right, *with no mistakes*, the *first* time. It had to be 2,000 miles of perfection. The cable had a copper wire core as one conductor (see Figure 5.3.1) and the ocean itself as the "return," with an insulating coating separating the two to prevent short-circuiting. And then, around all of that, the cable was encased in a layer of steel armor wire. The crucial insulation was made from gutta-percha, a Malaysian tree gum known to English scientists since 1843.[16] With insulating properties better than rubber, it was a particularly attractive substance, because it was *thermoplastic*, which meant it could be manipulated at high temperature to coat the inner conductor, but would become a very hard solid when cooled. The cold depths of the Atlantic were a perfect home for gutta-percha and, until the introduction of polyethylene in 1933, it was used in nearly all marine telegraph cables after the Atlantic cable.

The Atlantic cable was a technological masterpiece, the result of a flawless amalgamation of science and engineering, combined with real-world capitalism and politics.[17]

5.4 Solving the Atlantic Cable Equation

Solving (5.3.3) for the voltage $v(x, t)$ on the Atlantic cable is clearly our next task, *a task we have actually already accomplished*. You'll recall from Section 4.3 that we solved the heat equation

$$\frac{\partial u}{\partial t} = k \frac{\partial^2 u}{\partial x^2}$$

under the conditions given in (4.3.1) and (4.3.2):

$$u(x, 0) = 0, \quad x > 0,$$

and

$$u(0, t) = U_0, \quad t \geq 0.$$

The solution, as given in (4.3.23), is

$$u(x, t) = U_0 \left[1 - \text{erf} \left(\frac{x}{2\sqrt{kt}} \right) \right].$$

But all of this also describes an initially "dead" Atlantic cable with temperature $u(x, t)$ replaced with the voltage $v(x, t)$, and the temperature U_0 replaced with a battery of voltage U_0 that has been suddenly connected (as by the closing of a telegraph key) to the $x = 0$ end of the cable. By convention, U_0 is generally taken to be 1 volt for all $t \geq 0$ (as you'll see in Chapter 6, this in no way will be a restrictive assumption). This particular input voltage is called by physicists and electrical engineers a *unit step voltage*, because if you plot

$$v(0, t) = \begin{cases} 0, & t < 0 \\ 1, & t > 0 \end{cases},$$

it *looks* like a sidewise view of a step.

So, just like that, because he had read *Analytical Theory*, Thomson had the voltage response of the Atlantic cable to a unit step voltage input. That is, since we've already argued that $k = 1/RC$,

$$(5.4.1) \qquad v(x, t) = 1 - \text{erf} \left(\frac{x}{2} \sqrt{\frac{RC}{t}} \right), \quad t \geq 0.$$

Recalling the definition of the error function from (4.3.21), we can write (5.4.1) as

$$(5.4.2) \qquad v(x,t) = 1 - \frac{2}{\sqrt{\pi}} \int_0^{\frac{x}{2}\sqrt{\frac{RC}{t}}} e^{-y^2}\, dy.$$

The reason (5.4.2) is useful is because when a signal was sent over the cable, that signal's *current* was the physical quantity detected, not the voltage.[18] To calculate the Atlantic cable current, we use (5.3.1), which now reads

$$i(x,t) = -\frac{1}{R}\frac{\partial v}{\partial x} = \frac{2}{R\sqrt{\pi}}\left\{\frac{\partial}{\partial x}\int_0^{\frac{x}{2}\sqrt{\frac{RC}{t}}} e^{-y^2}\, dy\right\}.$$

To perform the differentiation of the integral, we need to use Leibniz's formula (see the Appendix), which gives us

$$i(x,t) = \frac{2}{R\sqrt{\pi}}\left\{e^{-x^2RC/4t}\frac{d}{dx}\left(\frac{x}{2}\sqrt{\frac{RC}{t}}\right)\right\} = \left(\frac{2}{R\sqrt{\pi}}\right)\left(\frac{1}{2}\sqrt{\frac{RC}{t}}\right)e^{-x^2RC/4t},$$

or

$$(5.4.3) \qquad i(x,t) = \sqrt{\frac{C}{\pi Rt}}\, e^{-x^2RC/4t},\ t \ge 0,\ x \ge 0.$$

Figure 5.4.1 shows a plot of (5.4.3), for the values of $R = 3$ ohms/nm, $C = 0.5$ microfarads/nm, and $x = 2{,}000$ nautical miles (nm), for the first 10 seconds after the input voltage step is applied. The plot has three features of practical engineering significance. First, there is a significant delay of about 3 seconds in the occurrence of the peak current. (The fact that the current has a peak value is the second important feature.[19]) Since the receiver at the $x = 2{,}000$ nm end of the cable responds most strongly to this peak current, this delay introduces an upper limit on the transmission speed over the cable. And finally, Figure 5.4.1 shows that a 1-volt step input produces a maximum current that is slightly in excess of 80 microamperes. Thus, if the actual sending voltage was, say, 77 volts (the voltage used on the 1858 cable) there would be at $x = 2{,}000$ nm a maximum received current of more than 6 milliamperes, a current easily detectable by Thomson's marine galvanometer.

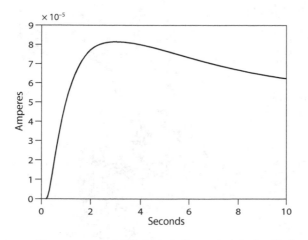

FIGURE 5.4.1. The current at $x = 2,000$ nautical miles on the Atlantic cable, in response to a unit step input voltage at $t = 0$.

We can find the time at which the peak current occurs by differentiating (5.4.3) with respect to time (and then the actual peak current by plugging that time back into (5.4.3)); if you do that, you'll find

$$(5.4.4) \qquad\qquad t_{max} = \frac{1}{2} RCx^2$$

and

$$(5.4.5) \qquad\qquad i_{max}\left(t_{max}\right) = \frac{1}{Rx} \sqrt{\frac{2}{\pi e}}.$$

Using the values in the text for R, C, and x, the results for t_{max} and i_{max} produced by (5.4.4) and (5.4.5), respectively, are in excellent agreement with Figure 5.4.1.

The result in (5.4.4), that t_{max} depends on x *squared*, became known as *Thomson's law of squares*, and it was the cause of much controversy and concern among those who considered investing money in the laying of really long submarine telegraph cables. According to (5.4.4), doubling

FIGURE 5.4.2. Father Neptune blesses Uncle Sam and Britannia on the success of the 1866 Trans-Atlantic telegraph cable. *Punch*, August 11, 1866.

the length of a cable would *quadruple* the time delay. In his 1855 paper, Thomson mentions that a 200-mile-long cable connecting Greenwich (England) and Brussels (Belgium) had a $\frac{1}{10}$ second delay, and so the law of squares predicts that "the retardation in a cable of [equal construction] extending half around the world (14,000 miles) would be $\left(\frac{14000}{200}\right)^2 \times \frac{1}{10} = 490$ seconds or $8\frac{1}{6}$ minutes." That's a long time to send a single character. That was a hypothetical calculation for a hypothetical cable, however, and the transmission speed on the 1866 Atlantic cable was actually a serviceable 6 to 8 words per minute.

That, in fact, was a huge advance over the performance of the 1858 cable, which actually worked for a number of weeks (failing, probably, because its gutta-percha insulation had been improperly deposited when the cable was being constructed). Before it failed, its apparent success was celebrated by an exchange of messages between Queen Victoria and President James Buchanan. The Queen's 98 word message took over 16 hours to send (!), and the President's 149 word reply took more than 10 hours to transmit. Why the dramatic difference in transmission speed over the same cable? Why should the direction of sending make any difference at all? And it was a *big* difference—a more than 50% increase in time to send a message a third shorter! This odd phenomenon, which had long puzzled telegraphists operating on shorter submarine cables, wasn't fully explained until more than a decade later. Think about this, and I'll show you, at the end of Chapter 6, how what appears to be a paradox was finally resolved in 1877. (The answer is not that one operator was simply faster on the key than was his counterpart! Instead the answer lies in a clever application of Kirchhoff's current law and the solution of a simple first-order, linear differential equation.)

At the failure of the 1858 cable, no less an authority than England's Astronomer Royal, the mathematician George Biddell Airy (1801–1892), thought to explain that failure by declaring the possibility of an Atlantic cable to be an absurdity, to be a virtually certain "mathematical impossibility." (That argument oddly ignored the fact that the cable *had* worked, if only for a while.) And yet, in the summer of 1866, there it was, clicking merrily away and making a lot of money for its investors. As Figure 5.4.2 shows, the success of the 1866 Atlantic cable was received, on both sides of the ocean, with great enthusiasm—along with a touch of apprehension at what it might mean for the future.

CHAPTER 6
Epilogue

6.1 What Came after the 1866 Cable

After Roger Bannister (1929–2018) ran the first sub–4 minute mile, a physiological feat long thought to be an impossibility, it seemed that people immediately started doing it on a regular basis. It was the same for the construction of long submarine telegraph cables. The same year that saw the Atlantic cable saw the 90 miles of ocean from Florida to Cuba conquered, and that didn't seem as amazing as it might have just a year earlier. Eight years later, Portugal was connected to Brazil with a cable longer (nearly 3,400 miles) than the 1866 Atlantic cable. Over the next three decades, similar cables would link the British Empire to places once thought by inhabitants of London to be nearly as remote as the far side of the Moon, including the Far East and Australia. Transmission speeds slowly increased, as well, and by 1898, it was common to see cables operating at 40 words per minute.

With all this success, however, it should be noted that one of the original worries about the Atlantic cable proved to be correct. Humans *could*, indeed, find suspect uses for the new technological wonder! A classic example of this occurred just a bit more than a year after the cable began operation. In November 1867, the English novelist Charles Dickens sailed to New England, to begin an American reading tour. The big issue on his mind didn't concern literary matters, though, but rather he was focused on whether or not it would be "safe" to have his mistress sail from England to join him. Once in Boston he quickly realized that the press would report every coming-and-going of a world-famous celebrity (reports that would certainly come to the attention of his wife), and so he decided to cable his personal assistant back in London, using a pre-arranged code, to cancel the lady's ocean trip.[1] It wasn't trivial for

Dickens, of course, but it certainly wasn't an application the creators of the cable would have used to interest potential investors.

The application of telegraphy in affairs of the heart appeared in a more uplifting form with the 1880 publication of a charming novella by the former American telegraph operator Ella Cheever Thayer (1849–1925). *Wired Love: A Romance of Dots and Dashes*, is the story of two telegraphers who have never met, and it reads today like a 19th-century prediction of the modern Internet chat room. The success of the Atlantic cable 14 years earlier surely increased the book's potential audience, and it was a best seller for years.

Well, okay, historical look-back *is* fun to do, but this is quite enough of that, and so back to mathematical physics and engineering!

The technological development of submarine cables after 1866 rode on the back of increasing theoretical understanding. That the retardation of the maximum signal current on the Atlantic cable depends on *x squared* was a surprising result and, indeed, not everybody believed it. As one British trade journal (*Electrical World*) recalled, in an 1892 article, "Thomson's prediction of the retardation that would be experienced struck practical electricians aghast. They were very much inclined not to believe in this mathematical newcomer."

Some felt the retardation might simply be the result of assuming the input was a step voltage. Perhaps other, more "real-life" signals (for example, finite-duration *pulses*, as would occur in sending a telegraph message via Morse code) would produce a different result. This ignored the fact that in his 1855 paper (note 15 in Chapter 5) Thomson had not limited himself to step inputs; he had also analyzed the cable current for a pulsed input signal of duration T. He had then let $T \to 0$ in such a way that a fixed value of charge was always injected into the $x = 0$ end (that is, as $T \to 0$, the pulse amplitude increased).[2] Even for this signal, however, he found "the time at which the maximum electrodynamic effect of connecting the battery for [just] an instant" is

$$t_{max} = \tfrac{1}{6} RCx^2,$$

which, with a minor change in the multiplicative factor from that in (5.4.4), is again a square law.

Despite the fact that there is a nonzero current well before 3 seconds, the *peak* current in Figure 5.4.1 occurring at a distance of 2,000 miles in 3 seconds was interpreted by many to mean electricity in the cable traveled at about 700 miles per second (recall, in contrast, our calculation of the electron *drift* speed in Section 5.2), and this belief persisted for a long time. Thomson himself might have been responsible for this error. In an 1860 encyclopedia essay titled "Velocity of Electricity," he started by listing the widely varying experimental values that had been observed using cables of different lengths, and then stated "now it is obvious, from the results which have been quoted, that the supposed 'velocity' of transmission of electric signals is not a definite constant."

And then, at the end of his essay, he wrote, "retardations [are] proportional to the squares of the distances travelled. . . . In other words, the 'velocity' of propagation might be said to be inversely proportional to the distance travelled." Thomson's use of the word *retardations* shows a proper understanding of its distinction from velocity, but then he did use the word *velocity* (even if in scare quotes), which is all that many remembered.

Another basic result that came from Thomson's analysis, besides the law of squares, had to do with what modern electrical engineers call *dispersive phase distortion*. If we resolve the cable input signal into its Fourier series representation as a sum of harmonically related sinusoidal functions (or a Fourier integral, again involving sinusoidal functions with *all* frequencies present), then each of those frequency components—which form what is called the signal's *spectrum*—would diffuse into the cable at a different speed. (We'll develop the diffusion speed in just a moment.) This results in a spreading in time of the signal, and so what might start off as a short-duration, sharply defined pulse at the transmitter key could easily arrive at the receiver as a weak, smeared-out, hard-to-detect dribble. Here's how a modern analysis (due to Thomson's friend, Lord Rayleigh) shows exactly what happens.

We begin by writing (with the aid of Euler's fabulous formula) the cable input as the real part of a complex-valued sinusoidal signal:

$$(6.1.1) \qquad v(0, t) = \text{Re}\{e^{i\omega t}\} = \cos(\omega t),$$

where ω is the radian frequency of some particular signal component ($\omega = 2\pi f$, where f is the frequency in the unit that used to be called *cycles per second* and is now called *hertz* (Hz), after the German mathematical physicist Heinrich Hertz (1857–1894)).[3] If we assume that the time variation of $v(x,t)$ everywhere along the cable is of the form $e^{i\omega t}$, and that the spatial variation can be separated out in product form, that is,

$$(6.1.2) \qquad v(x,\ t) = A(x)e^{i\omega t},\ A(0) = 1,$$

then substitution of (6.1.2) into the cable equation (5.3.3) gives

$$\frac{1}{RC}\frac{d^2A}{dx^2}e^{i\omega t} = Ai\omega e^{i\omega t},$$

or

$$(6.1.3) \qquad \frac{d^2A}{dx^2} = i\omega RCA.$$

We solve (6.1.3) by making the usual assumption of an exponential solution. That is, we assume

$$(6.1.4) \qquad A(x) = e^{px},$$

where p is some constant. Then, putting (6.1.4) into (6.1.3) gives

$$p^2 e^{px} = i\omega RCe^{px},$$

or

$$p = \pm\sqrt{i\omega RC} = \pm\sqrt{\omega RC}\sqrt{i}.$$

Since Euler's identity tells us that $i = e^{i\frac{\pi}{2}}$, then

$$\sqrt{i} = e^{i\frac{\pi}{4}} = \cos\left(\frac{\pi}{4}\right) + i\sin\left(\frac{\pi}{4}\right) = \frac{1}{\sqrt{2}} + i\frac{1}{\sqrt{2}},$$

and so

$$p = \pm \left[\sqrt{\frac{\omega RC}{2}} + i \sqrt{\frac{\omega RC}{2}} \right].$$

Only the negative root for p makes physical sense on a very long (infinitely long) cable, because the root with a positive real part would result in (6.1.4) giving an unbounded $v(x, t)$. Thus, $v(x, t)$ is the real part of

$$e^{-\left[\sqrt{\frac{\omega RC}{2}} + i \sqrt{\frac{\omega RC}{2}} \right] x} e^{i\omega t} = e^{-x\sqrt{\frac{\omega RC}{2}}} e^{-ix\sqrt{\frac{\omega RC}{2}}} e^{i\omega t} = e^{-x\sqrt{\frac{\omega RC}{2}}} e^{i\left[\omega t - x\sqrt{\frac{\omega RC}{2}} \right]}$$

and so

$$(6.1.5) \quad v(x, t) = e^{-x\sqrt{\frac{\omega RC}{2}}} \cos\left(\omega t - x\sqrt{\frac{\omega RC}{2}} \right), \; t \geq 0, \; x \geq 0.$$

From (6.1.5), we see that the cable voltage, in response to a sinusoidal input, is also a sinusoid, but with an amplitude that decays exponentially with distance from the input. The higher the frequency (ω), the more rapid is the decay.

To understand what is meant by *the speed of diffusion*, imagine a surfer riding the voltage signal into the cable at a speed such that she observes only the exponential amplitude decay, and *not* the sinusoidal time behavior. This is what a real surfer sees as she rides a water wave, seeing only a decline in water height, but no change in wave shape. This condition requires that the value of the cosine in (6.1.5) remain constant, and so the surfer's speed is determined by

$$\omega t - x\sqrt{\frac{\omega RC}{2}} = \text{constant},$$

which when differentiated, says

$$\omega - \frac{dx}{dt}\sqrt{\frac{\omega RC}{2}} = 0,$$

or

$$(6.1.6) \qquad \frac{dx}{dt} = \sqrt{\frac{2\omega}{RC}}.$$

The higher the frequency, the greater the diffusion speed is, but as we just argued, the greater, too, is the amplitude decay. In general, high-frequency signals on the Atlantic cable lived a fast life but died quickly, and so offered an electrical morality lesson that probably appealed to the Victorian idealization of rectitude.

For the Atlantic cable, $RC \approx 5 \times 10^{-17}$ s/cm² (see note 14 in Chapter 5) in cgs units, and so (6.1.5) says that for the amplitude of the signal at frequency ω to decay by a factor of e, the signal has to diffuse into the cable a distance of

$$x = \sqrt{\frac{2}{\omega RC}} = \sqrt{\frac{2}{2\pi f 5 \times 10^{-17}}} \text{cm} = \frac{431}{\sqrt{f}} \text{ nautical miles.}$$

Thus, a 100 Hz frequency component would decay by a factor of e every 43 nautical miles. In a 1,700 nautical-mile cable there would be a forty-fold effect, and a 1 volt input sine wave would emerge at the receiving end as less than 10^{-17} volts! A 10 Hz frequency component, however, would have a decay distance (by a factor of e) of 136 nautical miles, and a 1 volt sine wave would arrive as 3.7×10^{-6} volts, *several hundred billion* times larger than the 100 Hz component!

The diffusion speed in (6.1.6) works out to be

$$\frac{dx}{dt} = \sqrt{\frac{2\omega}{RC}} = \sqrt{\frac{2(2\pi f)}{5 \times 10^{-17}}} \frac{\text{cm}}{\text{s}} = 2,700\sqrt{f} \text{ nautical miles per second.}$$

Therefore, a 100 Hz frequency component would travel at 27,000 nautical miles per second, while the 10 Hz component loafed along at only 8,538 nautical miles per second, and this is the reason a signal with a broad frequency spectrum was pulled apart on the Atlantic cable.[4]

True understanding of the physics of the transmission of signals on cables didn't come until the crucial role of induction was appreciated. Induction was purposely left out of Thomson's 1855 analysis (see note 15 in Chapter 5 again), and its absence resulted in some puzzles that greatly bothered analysts. One you may have already noticed in our earlier work in Chapter 4, on heat energy propagation, is that the heat equation allows instantaneous action at an infinite distance—look, for example, at (4.3.23), which gives the temperature $u(x, t)$ anywhere along an infinite rod. Even though the temperature is initially zero for all x, if the $x = 0$ end is then suddenly raised to temperature U_0, we see that (4.3.23) says $u(x, t) > 0$, no matter how large x is or how small t is. That is, the heat equation places no limit on the speed of heat-energy propagation, and so the heat equation violates special relativity.

Of course, in 1855 and for many years after, this issue didn't particularly bother the Victorians, since special relativity didn't arrive on the scene until 1905. Still, *infinite* speed did seem to be a bit odd, even in the 1800s! This feature of Fourier's heat equation appears in the Atlantic cable equation, too, of course, as inspection of (5.4.1) and (5.4.3) shows. Specifically, the expression for the diffusion speed on the cable indicates that if $f > 3,590$ Hz, then the diffusion speed exceeds that of light, contrary to Einstein. The way to eliminate such superrapid action, as well as to tremendously improve the practical performance of long submarine cables, is to introduce the effects of induction. This was done by the English mathematical electrical engineer (and former telegraph operator) Oliver Heaviside (1850–1925), but not without a monumental war of words with "established authority" on the pages of the technical journals of the day.[5]

In August 1876, Heaviside showed that an improvement on Thomson's 1855 analysis for a cable includes, in addition to the parameters R and C that were discussed in Section 5.3, a third parameter:

G = leakage conductance—the reciprocal of the leakage resistance per unit length (due to the imperfection of the cable insulation).

Then, in June 1887, Heaviside finally added the missing fourth parameter of

L = inductance per unit length (due to the magnetic field of the cable current).

This last parameter results *not* in the heat equation, but rather in what is called the *wave equation*. The wave equation is a second-order partial differential equation like the heat equation, but it contains a new term of the form $\frac{\partial^2 v}{\partial t^2}$. With the wave equation in hand, Heaviside's work showed that, if the values of R, C, G, and L are properly adjusted, the *distortionless* propagation of a signal can be achieved. That is, the shape of the signal is not degraded as it propagates along the cable, because, unlike in the dispersive Atlantic cable, a distortionless cable propagates each frequency component of a signal at the same speed.

Heaviside claimed he understood this by 1875, when he wrote (in 1897 in his *Electromagnetic Theory*), "I set my self [this problem] 22 years ago, when I first realized as a consequence of Maxwell's theory of self-induction, combined with W. Thomson's theory of the electric telegraph, that all disturbances travelled at finite speed." He added,

> All diffusion formulae (as in heat conduction) show instantaneous action to an infinite distance of a source, though only to an infinitesimal extent. It is a general mathematical property, but should be taken with salt in making applications to physics. To make the theory of heat diffusion be rational as well as practical, some modification of the equations is needed to remove the instantaneity, however little difference it may make quantitatively, in general.

The introduction of induction does this, and allows a cable to be operated at much higher speeds than before: by 1928, cable speeds had routinely reached 400 words per minute.[6]

To understand what all that means, consider Figure 6.1.1, which shows Heaviside's elaboration of the simple Atlantic cable model analyzed by

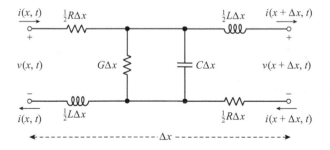

FIGURE 6.1.1. An arbitrary, very short section of Heaviside's cable.

Thomson (look back at Figure 5.3.1). Now, in addition to the per unit length resistance R and capacitance C, Heaviside included the additional per unit length inductance L and leakage conductance G (the reciprocal of the leakage resistance). As done in Figure 5.3.1, the inductance is inserted to make the cable symmetrical (that is, to look the same from left-to-right and right-to-left), and this is done by placing half the cable inductance in the top conductor on the right and the other half of the cable inductance in the bottom conductor on the left. The cable capacitance C is now in parallel with G.

Starting at the upper-left point of Figure 6.1.1 (where the current $i(x, t)$ enters the upper-left $\frac{1}{2}R\Delta x$ resistance) and summing voltage drops around the outer closed loop of the circuit, Kirchhoff's voltage law says

$$(6.1.7) \quad i(x,t)\frac{1}{2}R\Delta x + \frac{1}{2}L\Delta x \frac{\partial i(x+\Delta x, t)}{\partial t} + v(x+\Delta x, t)$$
$$+ i(x+\Delta x, t)\frac{1}{2}R\Delta x + \frac{1}{2}L\Delta x \frac{\partial i(x,t)}{\partial t} - v(x,t) = 0,$$

where partial derivatives are used because we have two independent variables (x and t). With a bit of rearranging, (6.1.7) becomes

$$(6.1.8) \quad -[v(x+\Delta x, t) - v(x,t)] = \frac{1}{2}R\Delta x[i(x,t) + i(x+\Delta x, t)]$$
$$+ \frac{1}{2}L\Delta x \frac{\partial}{\partial t}[i(x+\Delta x, t) + i(x,t)].$$

Then, dividing through (6.1.8) by Δx, we have

$$-\frac{[v(x+\Delta x, t) - v(x,t)]}{\Delta x} = \frac{1}{2}R[i(x,t) + i(x+\Delta x, t)]$$
$$+ \frac{1}{2}L\frac{\partial}{\partial t}[i(x+\Delta x, t) + i(x,t)],$$

and so, if we let $\Delta x \to 0$, we have $i(x+\Delta x, t) \to i(x, t)$, and we arrive at

$$(6.1.9) \qquad\qquad -\frac{\partial v}{\partial x} = iR + L\frac{\partial i}{\partial t},$$

which reduces to (5.3.1) if $L=0$ (the no-induction assumption of Thomson's Atlantic cable analysis).

Once we have (6.1.9) down on paper, we realize (perhaps) that we should have been able to skip the math and to have argued directly that the loss of voltage (the meaning of the minus sign) along the line per unit length is equal to the ohmic voltage drop per unit length plus the inductive voltage drop per unit length. With that physical insight as inspiration, we can immediately write a similar equation for the loss of current along the line per unit length as equal to the leakage current between the two conductors per unit length plus the current needed to charge the line's capacitance between the two conductors per unit length. That is,

$$(6.1.10) \qquad -\frac{\partial i}{\partial x} = Gv + C\frac{\partial v}{\partial t},$$

which reduces to (5.3.2) if $G=0$ (the no-leakage assumption of Thomson's Atlantic cable analysis).

If we differentiate (partially) our two equations, (6.1.9) with respect to x, and (6.1.10) with respect to t, we have

$$(6.1.11) \qquad -\frac{\partial^2 v}{\partial x^2} = R\frac{\partial i}{\partial x} + L\frac{\partial^2 i}{\partial x \partial t}$$

and

$$(6.1.12) \qquad -\frac{\partial^2 i}{\partial t \partial x} = G\frac{\partial v}{\partial t} + C\frac{\partial^2 v}{\partial t^2}.$$

Since x and t are independent, we can plausibly assume that the order of differentiation doesn't matter; that is,

$$\frac{\partial^2 i}{\partial x \partial t} = \frac{\partial^2 i}{\partial t \partial x}.$$

Thus, putting (6.1.10) and (6.1.12) into (6.1.11), we arrive at

$$-\frac{\partial^2 v}{\partial x^2} = R\left[-Gv - C\frac{\partial v}{\partial t}\right] + L\left[-G\frac{\partial v}{\partial t} - C\frac{\partial^2 v}{\partial t^2}\right],$$

which is a second-order partial differential equation called the *first telegraphy equation*:

(6.1.13) $$\frac{\partial^2 v}{\partial x^2} = LC\frac{\partial^2 v}{\partial t^2} + (RC + LG)\frac{\partial v}{\partial t} + RGv.$$

In much the same way, you can alternatively eliminate v to get a second-order partial differential equation for i, known as the *second telegraphy equation*:

(6.1.14) $$\frac{\partial^2 i}{\partial x^2} = LC\frac{\partial^2 i}{\partial t^2} + (RC + LG)\frac{\partial i}{\partial t} + RGi.$$

The structural similarity of the two telegraphy equations is striking.

There are several special cases of the telegraphy equations that are of particular interest. For example, the case of the trans-Atlantic cable is what electrical engineers call *a leakage-free* ($G = 0$), *noninductive* ($L = 0$) cable. Under those assumptions, (6.1.13) reduces to

(6.1.15) $$\frac{\partial^2 v}{\partial x^2} = RC\frac{\partial v}{\partial t},$$

which is, of course, the heat equation (5.3.3). Another situation, called the *lossless cable*, assumes $R = G = 0$ (*lossless*, because there are no ohmic energy dissipation mechanisms present). In that case, (6.1.13) reduces to

(6.1.16) $$\frac{\partial^2 v}{\partial x^2} = LC\frac{\partial^2 v}{\partial t^2} = \frac{1}{c^2}\frac{\partial^2 v}{\partial t^2},$$

where the constant $c = \dfrac{1}{\sqrt{LC}}$ has a physically important interpretation (to be revealed in just a moment). Equation (6.1.16) is called the *wave equation* (you'll see why, soon), and it was solved years before Bernoulli introduced separation of variables to solve the heat equation (look back at note 1 in Chapter 4). What follows is the very clever idea (dating from 1746) due to the French mathematical physicist Jean Le Rond d'Alembert (1717–1783). Although this idea doesn't have the wide applicability of Bernoulli's separation of variables method, it *is* nontheless fiendishly ingenious and well worth the 5 minutes (from start to finish) it will take you to absorb.

We start by making the change of variables

$$(6.1.17) \qquad\qquad r = x + ct$$

and

$$(6.1.18) \qquad\qquad s = x - ct.$$

Then, from the chain rule of freshman calculus, we have

$$(6.1.19) \qquad \frac{\partial v}{\partial x} = \frac{\partial v}{\partial r}\frac{\partial r}{\partial x} + \frac{\partial v}{\partial s}\frac{\partial s}{\partial x} = \frac{\partial v}{\partial r} + \frac{\partial v}{\partial s},$$

because

$$\frac{\partial r}{\partial x} = \frac{\partial s}{\partial x} = 1.$$

In the same way,

$$(6.1.20) \qquad \frac{\partial v}{\partial t} = \frac{\partial v}{\partial r}\frac{\partial r}{\partial t} + \frac{\partial v}{\partial s}\frac{\partial s}{\partial t} = c\frac{\partial v}{\partial r} - c\frac{\partial v}{\partial s},$$

because

$$\frac{\partial r}{\partial t} = c, \quad \frac{\partial s}{\partial t} = -c.$$

Next, we apply the chain rule again, to write

$$\frac{\partial^2 v}{\partial x^2} = \frac{\partial}{\partial x}\left(\frac{\partial v}{\partial x}\right) = \frac{\partial}{\partial r}\left(\frac{\partial v}{\partial x}\right)\frac{\partial r}{\partial x} + \frac{\partial}{\partial s}\left(\frac{\partial v}{\partial x}\right)\frac{\partial s}{\partial x},$$

which is, using (6.1.19) for $\frac{\partial v}{\partial x}$ and remembering that $\frac{\partial r}{\partial x} = \frac{\partial s}{\partial x} = 1$,

$$\frac{\partial^2 v}{\partial x^2} = \frac{\partial}{\partial r}\left(\frac{\partial v}{\partial r} + \frac{\partial v}{\partial s}\right) + \frac{\partial}{\partial s}\left(\frac{\partial v}{\partial r} + \frac{\partial v}{\partial s}\right),$$

or

(6.1.21)
$$\frac{\partial^2 v}{\partial x^2} = \frac{\partial^2 v}{\partial r^2} + 2\frac{\partial^2 v}{\partial r\,\partial s} + \frac{\partial^2 v}{\partial s^2}$$

if we assume $\frac{\partial^2 v}{\partial r\,\partial s} = \frac{\partial^2 v}{\partial s\,\partial r}$. In the same way (which I'll let you confirm),

(6.1.22)
$$\frac{\partial^2 v}{\partial t^2} = c^2\left(\frac{\partial^2 v}{\partial r^2} - 2\frac{\partial^2 v}{\partial r\,\partial s} + \frac{\partial^2 v}{\partial s^2}\right).$$

If we substitute (6.1.21) and (6.1.22) into (6.1.16), we have

$$\frac{\partial^2 v}{\partial r^2} + 2\frac{\partial^2 v}{\partial r\,\partial s} + \frac{\partial^2 v}{\partial s^2} = \frac{1}{c^2}\left[c^2\left(\frac{\partial^2 v}{\partial r^2} - 2\frac{\partial^2 v}{\partial r\,\partial s} + \frac{\partial^2 v}{\partial s^2}\right)\right],$$

which reduces to

(6.1.23)
$$\frac{\partial^2 v}{\partial r\,\partial s} = 0.$$

If we then integrate (6.1.23) twice, once with respect to s (to get $f(r)$, an arbitrary function of r) and once with respect to r (to get $g(s)$, an arbitrary function of s), we have

$$v(r, s) = f(r) + g(s),$$

or returning to our original variables of x and t,

(6.1.24) $$v(x, t) = f(x + ct) + g(x - ct).$$

The functions f and g each have interesting physical interpretations. Each is a *wave* traveling at speed c, but in opposite directions, along the cable. To see this, focus on $g(x - ct)$. At $t = 0$, this is the voltage distribution along the cable given by $g(x)$ and then, as t increases, each point on the voltage distribution is found to be shifting to the right (the direction of increasing x) if we keep the argument $x - ct$ constant. That is, if

$$x - ct = \text{constant},$$

then we see the distribution moving at the speed

$$\frac{dx}{dt} = c.$$

So, $g(x - ct)$ is a voltage distribution traveling at speed c in the direction of increasing x.

In the same way, $f(x + ct)$ is a voltage distribution traveling at speed c in the direction of decreasing x. The total solution to the lossless wave equation is the sum of these two *traveling wave solutions*. The specific nature of $f(x)$ and $g(x)$ is determined by the initial and boundary conditions for the cable. What happens if the cable is, more realistically, *not* lossless?

A more sophisticated analysis than I've done here shows, not surprisingly, that then the traveling voltage distributions f and g decay in amplitude with increasing time. What is surprising, however, was Heaviside's discovery (in his 1887 paper) that if the condition of (5.2.6) is satisfied, then the original voltage-distribution shape of a signal is preserved, *even in the presence of energy loss*. That is, each frequency component of the original voltage distribution travels at the same speed even as it decays in amplitude, and so a signal can be sent over such an adjusted cable without suffering distortion (the adjustment was in the form of adding inductance at spaced intervals to the cable, to achieve

(5.2.6), a process called *loading the cable*.[7] Note, carefully, for the assumed case of the Atlantic cable ($G = 0$ and $L = 0$) that (5.2.6) was most definitely *not* satisfied by that cable, and so it was *not* free of distortion.

Cables played an increasingly important role in world affairs as well, one that transcended the "merely" technical. It was the destruction of multiple trans-oceanic cables, for example, that led to a pivotal event in World War I when (in August 1914) England cut the five German cables that ran through the English Channel. Eventually Germany lost all of her access to trans-oceanic cables, and thereafter was able to communicate in real time with her diplomats in overseas postings strictly by radio. That opened up the possibility of message interception, and that's exactly what happened in the infamous affair of the Zimmermann telegram.

In early 1917, when America was still dithering over the issue of entering the war that had then been raging in Europe for 3 years, the German Foreign Secretary Arthur Zimmermann sent a coded radio message to the Imperial German Minister in Mexico, directing him to approach the Mexican government with an astounding proposal. In an effort to encourage a continuation of President Wilson's position of neutrality, Germany invited Mexico to join with her (and Japan, too, then fighting against Germany) in an invasion of America! In exchange for providing this "diversion" of America from the European conflict, Mexico would get back her lost territories of Texas, New Mexico, and Arizona. The British intercepted this ludicrous message (one that reads today like it came from a Monty Python movie script) and leaked[8] it to Washington— where it caused sufficient outrage that America soon after declared war on Germany.

If the trans-oceanic cables had not been cut, if the radio-transmitted Zimmermann telegram had not been intercepted and leaked, would America have sat out World War I? Well, who knows? Historians continue to debate this question of alternative history to this day. But one cannot deny the claim that the cutting of the German cables did play a central role in the eventual involvement of America in that awful conflict.

Submarine cables are in operation to this day. (Their inherent remoteness offers highly secure communication channels that, unlike the radio transmissions of the Zimmermann affair, are very difficult to hack into—

but not impossible. In the early 1970s, an American submarine success-fully tapped a Soviet military telephone cable 400 feet deep in the Sea of Okhotsk, and later another sub repeated that success in the Barents Sea.) The technology of submarine cables eventually left simple Morse code telegraphy, from the days of the Atlantic cable, in the past. The first telephone voice circuit went into trans-Atlantic operation in 1956 (a sub-stantially shorter submarine voice cable actually started operating be-tween Key West, Florida, and Havana, Cuba, much earlier, in 1921), and today the floor of the Atlantic Ocean is home to a vast network of fiber-optic digital data cables operating at many hundreds of gigabits per sec-ond, carrying text, voice, and television signals protected by automatic error detection and correction coding. The Atlantic cable pioneers would be mightily impressed, perhaps even to the extent as to wonder whether maybe there was just a bit of magic, too, mixed in with the science and engineering.

6.2 The Cable Equation, Duhamel's Integral, and Electronic Computation

In this, the final section of the book, we'll straddle the past and the future. The past, in that we'll deal with the Atlantic cable heat diffusion equation as given by (5.3.3), and the future in that I'll show you how, using modern computational tools, we can solve the Atlantic cable equation for input signals that would have left Thomson dumbstruck. As an example of such a signal, suppose for the cable input voltage we have

$$(6.2.1) \qquad v(0, t) = \sin\{t\sin\{t\sin\{t\}\}\}, \quad 0 \leq t \leq 5,$$

and zero otherwise. What is the cable voltage at $x = 2,000$ nm for the case of $R = 3$ ohms/nm and $C = 0.5$ microfarads/nm? One look at a plot of (6.2.1), in Figure 6.2.1, would have turned even a math whiz like Thom-son pale. (There is no special significance to this signal, other than that it looks pretty complicated.)

This prompts the following problem. While the response—given in (5.4.2)—of the Atlantic cable to an input of the unit step voltage is

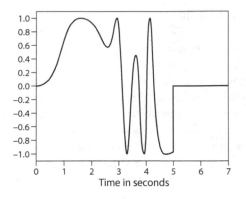

FIGURE 6.2.1. The cable input signal $v(0, t) = \sin\{t\sin\{t\sin\{t\}\}\}$, $0 \leq t \leq 5$.

one of obvious interest to a telegraph operator who has just closed his key, there are many other potentially interesting input signals besides (6.2.1) that one can easily imagine. Do we have to do a separate analysis for each of them, to get the associated response? The happy answer is *no*. Once we know the cable's response to the unit step voltage, we can then find the cable's response to *any other* input signal. What we are going to do, then, to find the cable voltage in response to (6.2.1), is develop a way to find the cable voltage in response to any input signal in terms of the cable's response to a unit step. Then, the answer to our problem will just be a special case of that general solution.

Learning how to do this is one of the central tasks of earning an undergraduate degree in electrical engineering, and that might seem to imply that a long, arduous journey lies ahead. *Au contraire!* You're going to learn how to do this in just *the next few pages*, and once you have finished those pages, you'll be able to answer questions to which both Fourier and Thomson, brilliant as they were, would have thrown up their hands in defeat. One reason we'll be able to do this in such an abbreviated space is that I'm going to make two simplifying assumptions, one about the nature of the input and the other about the behavior of the cable. Those assumptions are actually quite weak, and in fact, both can be totally eliminated, but they will let us immediately sidestep all manner of irritating, pesky technical issues that would add little to the discussion here of solving the Atlantic cable equation for an arbitrary input.

First, if we call the $x = 0$ input signal $e(t)$, $t \geq 0$, then I'm going to assume that

(6.2.2) $e(0) = v(0, 0) = 0.$

You'll notice that (6.2.1) satisfies (6.2.2). From a practical viewpoint, this isn't really a very restrictive assumption, as every signal is zero until we turn it on, and nothing in the real world changes instantly. So, if $e(t)$ is zero just before (written as time $t = 0-$) we apply it to the cable input, then it will still be zero *at* $t = 0$. And if you still find this too much to accept, well then, just pretend my $e(t)$ is your input signal $\hat{e}(t)$—where $\hat{e}(0) \neq 0$—and I'll write my $e(t)$ as

$$e(t) = \hat{e}(t)[1 - e^{-10^{100}t}].$$

You'll notice that, irrespective of what your $\hat{e}(0)$ might be, my $e(0) = 0$ *and* my $e(t) = \hat{e}(t)$ for all t greater than, say, 10^{-99} seconds. So, we are *both* happy!

The second assumption I'm going to ask you to accept is a plausible one, but it actually (to really justify it) requires more technical discussion than I'm going to give it. Instead, I'm going to fall back on your willingness to live with the plausibility of it, but you should realize that I am being just a bit casual here. Suppose we apply a unit step voltage to the cable input at time $t = 0$ and then, $k\Delta\tau$ seconds later (where k is any positive integer, and $\Delta\tau$ is some small increment of time), we apply another unit step voltage to the cable input. If the response to the unit step applied at $t = 0$ is $H(t)$, then the response to the later unit step is the delayed response $H(t - k\Delta\tau)$, and the total response to both inputs is $H(t) + H(t - k\Delta\tau)$. This is true if the cable initially has zero stored energy in it (as electrical engineers like to put it, if the cable is "dead"), and this behavior can be shown to be the result of the linearity of the circuit elements (resistors and capacitors) that are in the model for the Atlantic cable. This behavior is called *superposition*, and its assumption is the cornerstone of our analysis.

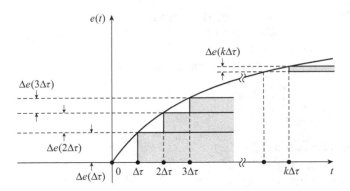

FIGURE 6.2.2. Staircase approximation to a smooth curve.

With the above in mind, Figure 6.2.2 shows an arbitrary input signal to the cable, $e(t)$, along with a stepwise approximation, with each new step occurring the same delay of $\Delta\tau$ seconds after the previous step. Speaking as engineers and physicists, we claim that as $\Delta\tau \to 0$, the approximation "gets better and better." A mathematician might cringe at this, but only just a little bit, because, while the wording might be rough and ready, the intuitive idea *can* be made pure. I'm going to assume that you can immediately appreciate the plausibility of Figure 6.2.2.

We know from (5.4.2) that a unit step applied to the cable input at time $t = 0$ produces the response

(6.2.3)
$$H(t) = \begin{cases} 1 - \dfrac{2}{\sqrt{\pi}} \displaystyle\int_0^{\frac{x}{2}\sqrt{\frac{RC}{t}}} e^{-y^2}\, dy, & t \geq 0 \\ 0, & t < 0 \end{cases}.$$

Now, concentrate on the typical step input starting at time $t = k\Delta\tau$, where k is any nonnegative integer, with amplitude $\Delta e(k\Delta\tau)$. By our second assumption, this step will produce an *amplitude-scaled* response of

$$\{\Delta e(k\Delta\tau)\} H(t - k\Delta\tau) = \left\{ \frac{\Delta e(k\Delta\tau)}{\Delta\tau} \right\} \Delta\tau H(t - k\Delta\tau).$$

If we add all the individual, delayed responses due to the individual steps, we will have, by superposition, the total response. If we call the result $r(t)$, then

$$r(t) \approx \sum_{k=0}^{t/\Delta\tau} \left\{ \frac{\Delta e(k\Delta\tau)}{\Delta\tau} \right\} \Delta\tau H(t - k\Delta\tau).$$

Now, as we let $\Delta\tau \to 0$, we have $k\Delta\tau \to \tau$, and we recognize (by definition of the derivative) that

$$\lim_{\Delta\tau \to 0} \frac{\Delta e(k\Delta\tau)}{\Delta\tau} = \frac{de(\tau)}{d\tau}.$$

Further, as $\Delta\tau \to d\tau$, we (again) have $k\Delta\tau \to \tau$, and the summation turns into an integral, and so (as $t/\Delta\tau \to \infty$ as $\Delta\tau \to 0$) we arrive at

$$r(t) = \int_0^\infty \frac{de(\tau)}{d\tau} H(t - \tau) d\tau.$$

In the older engineering literature, this was often called *Duhamel's integral*,[9] but the modern terminology is that it is a *superposition* or *convolution* integral. Notice, carefully, that the upper limit of infinity on the integral can be replaced with t because, once $\tau > t$, we have H in the integrand with a negative argument and we know, from (6.2.3), that the step response of the cable is zero for a negative argument. (Unless you believe in time machines, there can be no step response *before* the step is applied!) Thus, for $\tau > t$, there is no additional contribution to the value of the integral, and we have

(6.2.4) $$r(t) = \int_0^t \frac{de(\tau)}{d\tau} H(t - \tau) d\tau.$$

One technical problem with (6.2.4) is that it requires that we know the derivative of the input. This is not attractive—would *you* want to differentiate (6.2.1)? It would be far more convenient instead to work directly with $e(t)$, and the derivative of $H(t)$, which we can easily calculate from (6.2.3). We can get (6.2.4) into such a form by integration by parts.[10] If we let

$$u(\tau) = H(t - \tau)$$

and

$$dv = \frac{de(\tau)}{d\tau} d\tau = de(\tau),$$

then

$$\frac{du}{d\tau} = \frac{d}{d\tau} H(t - \tau)$$

and

$$v = e(\tau).$$

Thus, we have

$$r(t) = \left\{ e(\tau)H(t-\tau) \right\} \Big|_0^t - \int_0^t e(\tau)\frac{d}{d\tau}H(t-\tau)d\tau$$

$$= e(t)H(0) - e(0)H(t) - \int_0^t e(\tau)\frac{d}{d\tau}H(t-\tau)d\tau.$$

Since $e(0) = 0$ by assumption, and $H(0) = 0$ from (6.2.3), we have the response of the Atlantic cable to the *arbitrary* input $e(t)$ as

$$r(t) = -\int_0^t e(\tau)\frac{d}{d\tau}H(t-\tau)d\tau.$$

Next, change the variable to $s = t - \tau$. Then $ds = -d\tau$, and so

$$r(t) = -\int_t^0 e(t-s)\frac{d}{-ds}H(s)(-ds),$$

or

(6.2.5)
$$r(t) = \int_0^t e(t-s)\left\{\frac{d}{ds}H(s)\right\}ds.$$

From (6.2.3), we have

$$H(t) = 1 - \frac{2}{\sqrt{\pi}}\int_0^{\frac{x}{2}\sqrt{\frac{RC}{t}}} e^{-y^2}\,dy,$$

and so using Leibniz's formula (see the Appendix), we have

$$\frac{d}{dt}H(t) = -\frac{2}{\sqrt{\pi}}e^{-\frac{x^2RC}{4t}}\frac{x}{2}\sqrt{RC}\left(-\frac{1}{2}\frac{1}{t^{3/2}}\right) = \frac{x}{2}\sqrt{\frac{RC}{\pi}}\frac{e^{-\frac{x^2RC}{4t}}}{t^{3/2}},$$

or, making the (trivial) change from t to s,

$$\frac{d}{ds}H(s) = \frac{x}{2}\sqrt{\frac{RC}{\pi}}\frac{e^{-\frac{x^2RC}{4s}}}{s^{3/2}}.$$

Putting this into (6.2.5) gives us our general result for the cable voltage at distance x, at time t, with an input of $v(0, t) = e(t)$:

(6.2.6)
$$r(x,t) = \frac{x}{2}\sqrt{\frac{RC}{\pi}}\int_0^t e(t-s)\left\{\frac{e^{-\frac{x^2RC}{4s}}}{s^{3/2}}\right\}ds.$$

While (6.2.6) is a "pretty" result, it will only be useful if we can actually do the integral. When $e(t)$ is simple enough, (6.2.6) can be done analytically, but when I say *simple* I mean *really* simple. Certainly for the $e(t)$ of (6.2.1) and Figure 6.2.1, there is zero probability of evaluating (6.2.6) analytically. What I'll show you next is how to do that with the aid of an invention that Fourier surely never even dreamed of, and that William Thomson had but the faintest glimmer about: the modern, high-speed electronic computer (as I write—mid-2019—the world's fastest machine, at the Oak Ridge National Laboratory in Tennessee, can perform 200 quadrillion ($= 2 \times 10^{17}$) floating-point operations per second), in combination with powerful yet easy-to-use scientific programming software.

Fourier died far too early to have even imagined an electronic digital computer, but 15 years after the Atlantic cable went into operation, Thomson was actually involved in constructing a precursor to a digital computer—an *analog* computer that evaluated the partial sums of the Fourier series encountered in calculating the timing of tides. The Thomson machine used a flexible cord passing through a complicated arrangement of fixed and moveable pulleys. Alas, all the mechanical forces involved proved to be the inherent flaw of the device, as the resulting cord stretches limited the machine's accuracy to summing, at most, 15 terms.[11] Still, the *idea* of an automatic machine computing complicated mathematical expressions was in Thomson's mind as early as 1882, when he wrote, with some humor, "The object of this machine [the tidal harmonic analyzer] is to substitute brass for brains."[12]

The development of the electronic digital computer came decades after Thomson's death, along with the invention of the means for instructing such machines to carry out a desired task.[13] Those *codes* go by such names as Mathematica (popular with mathematicians) and MATLAB (popular with engineers). In many respects, however, these various codes (or *mathematical programming languages*, as they are commonly called) use much of the same syntax, and if you know one language, it is often not very difficult to decipher what a program written in another language is doing.

What I'm going to do next is first show you how easy it is to use MATLAB to answer the computational problem I challenged you with in note 14 of Chapter 4 (a problem that appears in Fourier's *Analytical Theory*). Then, second (to wrap up this book), I'll show you how to write the MATLAB code to evaluate the convolution integral in (6.2.6) using (6.2.1) as the input signal to the Atlantic cable. These two coding examples will be written in such a "high-level" way as to be understandable by anyone who can read English (that is not a metaphor or a joke, but the literal truth). In no way is this book meant to be a MATLAB primer, and so I'll say nothing about the nuances of MATLAB (and you don't need to know any of that, anyway, to understand what the codes are doing).

You'll recall the challenge question in Chapter 4 was to find the solutions to the equation

$$\lambda R = \tan(\lambda R),$$

and there I gave you the first four solutions for μ ($\lambda = \frac{\mu}{R}$), each to six decimal places. (R is the fixed radius of a sphere with its surface insulated.) If you look back at Figure 4.5.1, you should be able to convince yourself that for the first four solutions, beyond the trivial one of $\lambda R = 0$, we have

$$\pi < \lambda_1 R < \tfrac{3}{2}\pi,$$

$$2\pi < \lambda_2 R < \tfrac{5}{2}\pi,$$

$$3\pi < \lambda_3 R < \tfrac{7}{2}\pi,$$

$$4\pi < \lambda_4 R < \tfrac{9}{2}\pi,$$

and, for the calculation of the fifth solution (which is the challenge question),

$$5\pi < \lambda_5 R < \tfrac{11}{2}\pi.$$

Now, define the function

$$f(\lambda R) = \tan(\lambda R) - \lambda R.$$

We wish to find the value of λR such that $f(\lambda R) = 0$, where we know the fifth solution value of λR is somewhere in the interval $5\pi = 15.70796$ to $\frac{11}{2}\pi = 17.27875$. Notice that as we make λR larger and larger, $f(\lambda R)$

will become larger and larger (the tangent function monotonically increases faster than linearly as its argument increases). So, if we insert a value of λR greater than the solution value, we'll get $f(\lambda R) > 0$; and if we insert a value of λR smaller than the solution value, we'll get $f(\lambda R) < 0$. These two observations motivate the following procedure.

To start, assign the variable xl (for "x-lower,") a value known to be less than the solution value (say, $xl = 15.8$), and assign the variable xu (for "x-upper,") a value known to be greater than the solution value (say, $xu = 17.2787$). Thus, initially, $f(xu) > 0$, and $f(xl) < 0$. Next, calculate

$$x = \frac{1}{2}(xu + xl),$$

and then calculate the value of $f(x) = \tan(x) - x$. (The value of x is the value of λR.) If $f(x) > 0$, then x becomes the new value of xu. If, however, $f(x) < 0$, then x becomes the new value of xl. Then repeat this procedure, over and over, with each new repetition clearly cutting the length of the interval xl to xu (in which the solution lies) in half. It doesn't take long for this process to collapse xl to xu to a very narrow interval. After just 20 repetitions, for example, the length of the interval xl to xu has been reduced by a factor of $2^{20} > 1,000,000$. Computer scientists call this process, appropriately enough, the *binary chop algorithm*.

The MATLAB code **Lchop.m**—for "lambda chop"—implements this algorithm, and I think you can understand the code even if you've never before even heard of MATLAB. The first line sets xu and xl to their initial values. The next 10 lines form an endless (or infinite) *while* loop (*endless*, because 2 will always be greater than 0, or so we hope). Inside the loop, four things happen: (1) x is assigned the value midway between xu and xl; (2) the function $f(x) = \tan(x) - x$ is evaluated; (3) depending on the sign of $f(x)$, the value of x is assigned by an *if/else* structure to either xu or xl (the assumption here is that $f(x) = 0$ never occurs, but if this low-probability possibility worries you, you'll need to add some additional code); (4) and finally, the value of x is displayed on the computer screen. Then the entire process repeats, endlessly.

This all happens at high speed, and so the visual result on the computer screen is a fast, upward scrolling of the continually updated values of x, and you can watch as more and more of the digits of x stabilize, from left-to-right; very quickly, in fact, the screen appears not to be changing as all of the displayed digits reach their final values. On my 5-year-old, bottom-of-the-line laptop, this happens in less than a second with the display showing 15 decimal digits. While the display appears to be "frozen," in fact, **Lchop.m** is actually still running, and to exit the infinite *while* loop requires a keyboard entry of "Control-C." In any case, the answer to the challenge question is

$$\lambda_5 = \frac{\mu_5}{R} = \frac{17.220755}{R}.$$

```
%Lchop.m
xu=17.2787;xl=15.8;
while 2 > 0
    x=(xu+xl)/2;
    f=tan(x)-x;
    if f>0
        xu=x;
    else
        xl=x;
    end
    x
end
```

With this initial success in our pocket, let's now tackle the second computational problem I promised you we'd complete: how to use MATLAB to evaluate the behavior of the Atlantic cable in response to the input of (6.2.1). In fact, we are going to do just a bit more, in that the convolution integral of (6.2.6) is for the cable voltage. What would have really been of interest to Thomson, however, is not the voltage but rather the cable *current*, the physical quantity his marine galvanometer actually detected. So, instead of (6.2.6), which gives the cable voltage $r(x, t)$, we want to study the cable current $i(x, t)$ given by (5.3.1)—with $v(x, t)$ replaced by $r(x, t)$:

$$i(x,t) = -\frac{1}{R}\frac{\partial r}{\partial x} = -\frac{1}{R}\frac{\partial}{\partial x}\left[\frac{x}{2}\sqrt{\frac{RC}{\pi}}\int_0^t e(t-s)\left\{\frac{e^{-x^2RC/4s}}{s^{3/2}}\right\}ds\right].$$

That is,

$$i(x,t) = -\frac{1}{2}\sqrt{\frac{C}{\pi R}}\frac{\partial}{\partial x}\left[x\int_0^t e(t-s)\left\{\frac{e^{-x^2RC/4s}}{s^{3/2}}\right\}ds\right],$$

and so

$$i(x,t) = -\frac{1}{2}\sqrt{\frac{C}{\pi R}}$$

(6.2.7)
$$\times\left[\int_0^t e(t-s)\left\{\frac{e^{-x^2RC/4s}}{s^{3/2}}\right\}ds + x\frac{\partial}{\partial x}\int_0^t e(t-s)\left\{\frac{e^{-x^2RC/4s}}{s^{3/2}}\right\}ds\right].$$

From Leibniz's rule (see the Appendix), we have

$$\frac{\partial}{\partial x}\int_0^t e(t-s)\left\{\frac{e^{-\frac{x^2RC}{4s}}}{s^{3/2}}\right\}ds = \int_0^t e(t-s)\frac{\partial}{\partial x}\left\{\frac{e^{-\frac{x^2RC}{4s}}}{s^{3/2}}\right\}ds$$

$$= \int_0^t e(t-s)\left\{\frac{-2xRC}{4s}\frac{e^{-\frac{x^2RC}{4s}}}{s^{3/2}}\right\}ds$$

$$= -\frac{xRC}{2}\int_0^t e(t-s)\left\{\frac{e^{-\frac{x^2RC}{4s}}}{s^{5/2}}\right\}ds.$$

Thus,

$$i(x,t) = -\frac{1}{2}\sqrt{\frac{C}{\pi R}}$$

$$\times \left[\int_0^t e(t-s)\left\{ \frac{e^{-x^2RC/4s}}{s^{3/2}} \right\} ds - \frac{x^2RC}{2} \int_0^t e(t-s)\left\{ \frac{e^{-\frac{x^2RC}{4s}}}{s^{5/2}} \right\} ds \right],$$

or

(6.2.8) $$i(x,t) = \frac{x^2C}{4}\sqrt{\frac{RC}{\pi}} \int_0^t e(t-s)\left\{ \frac{e^{-\frac{x^2RC}{4s}}}{s^{5/2}} \right\} ds$$

$$-\frac{1}{2}\sqrt{\frac{C}{\pi R}} \int_0^t e(t-s)\left\{ \frac{e^{-x^2RC/4s}}{s^{3/2}} \right\} ds.$$

Wow! Those are pretty intimidating integrals once we put what $e(t)$ is from (6.2.1) into them. Fortunately, MATLAB can eat (6.2.8) for lunch. Here's how.

Any integral of the form

$$\int_0^t e(t-s)h(s)\,ds$$

is the *convolution* of $e(t)$ and $h(t)$, written as $e(t) \otimes h(t)$. So, defining

$$h_1(t) = \frac{e^{-\frac{x^2RC}{4t}}}{t^{5/2}}$$

and

$$h_2(t) = \frac{e^{-\frac{x^2RC}{4t}}}{t^{3/2}},$$

(6.2.8) becomes

$$(6.2.9) \qquad i(x,t) = \frac{x^2 C}{4}\sqrt{\frac{RC}{\pi}}\, e(t) \otimes h_1(t) - \frac{1}{2}\sqrt{\frac{C}{\pi R}}\, e(t) \otimes h_2(t).$$

MATLAB has a built-in command called *conv*, which accepts the two time functions $e(t)$ and $h(t)$ as arguments and computes the time function $e(t) \otimes h(t)$. So the numerical calculation of $e(t) \otimes h(t)$ becomes as simple to code as is $\sqrt{t}$. In fact, assuming $R = 3$ ohms/nm, $C = 0.5$ microfarads/nm, and $x = 2,000$ nautical miles (nm), the MATLAB code **fancyinput.m** does the job of computing and plotting (6.2.9) for the first 10 seconds after the input voltage $e(t)$ is defined and applied to the $x = 0$ end of the cable.

```
%fancyinput.m
01      tstart=1e-6;tstop=10;n=10000;x=2000;
02          R=3;C=5e-7;deltat=(tstop-tstart)/n;
03          t=tstart:deltat:tstop;
04          exponent=x*x*R*C/4;
05          f1=(x*x*C/4)*sqrt(R*C/pi);f2=sqrt(C/(pi*R))/2;
06          f3=exp(-exponent./t);
07          for k=1:length(t)
08              if t(k)<10
09                  input(k)=1;
10              else
11                  input(k)=0;
12              end
13          end
14          input=input*deltat;
15          h1=f3./(t.^2.5);h2=f3./(t.^1.5);
16          current=f1*conv(input,h1)-f2*conv(input,h2);
17          plot(t,current(1:length(t)),'-k')
18          xlabel('time (seconds)')
19          ylabel('cable current (amperes)')
```

Let me now walk you through **fancyinput.m** (the two-digit numbers at the far-left of each line of code are not part of MATLAB, but rather are simply reference tags that I have added so I can direct your attention to specific lines).

Line 01 defines the time span the code will be working in (the initial time is defined as 10^{-6} seconds, rather than zero, to avoid division-by-zero problems), the value of x, and the value of n (the number of discrete points in time the convolution integral evaluations will use). Since $n = 10,000$ for a 10-second time span, the individual time points are 0.001 seconds apart. Line 02 defines the values of R and C, and computes the value of the "differential" dt. Line 03 creates a *t-vector* of time points, starting from the initial time (10^{-6}) to the final time (10), in steps of dt. This vector has $n + 1$ elements $(10,001)$. Line 04 calculates the value of $\frac{x^2 RC}{4}$, and Line 05 calculates the values of $\frac{x^2 C}{4}\sqrt{\frac{RC}{\pi}}$ and $\frac{1}{2}\sqrt{\frac{C}{\pi R}}$. Line 06 calculates a *vector* of values $e^{-\frac{x^2 RC}{4s}}$, that is, a value for each point in the *t*-vector. Lines 07 through 13 calculate a vector of values for the input, with a value for each value in the *t*-vector. (The code in the shaded box is for the case of a unit step voltage, which is why all the values are equal to 1. The *length* command returns the number of elements in its vector argument.) Line 14 scales the input by dt. Line 15 calculates the vectors of values for $h_1(t)$ and $h_2(t)$, a value for each point in the *t*-vector. Line 16 implements (6.2.9), calculating a vector of values for the current. (This vector has $2n + 1$ elements $(20,001)$.) Line 17 creates a plot of current versus time. Lines 18 and 19 print the horizontal and vertical axis labels, respectively.

When executed, **fancyinput.m** produced Figure 6.2.3, showing the cable current at a distance of 2,000 nautical miles in response to a unit step voltage input. If you compare Figure 6.2.3 with Figure 5.4.1 (a plot of (5.4.3), the *exact analytical expression* for the current for a unit step input voltage), you'll see they are virtually identical. This serves as a good check on the correctness of the code.

To see what cable current results with the input defined by (6.2.1), all that has to be done is the replacement of Lines 07 through 13 with the following:

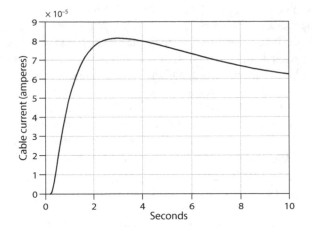

FIGURE 6.2.3. The output of **fancyinput.m** with a unit step input (compare with Figure 5.4.1).

```
for k=1:length(t)
    if t(k)<5
        input(k)=sin(t(k)*sin(t(k)*sin(t(k))));
    else
        input(k)=0;
    end
end
```

The result is Figure 6.2.4.

One thing that is immediately obvious, when Figure 6.2.4 is compared with Figure 6.2.1, is that the fine structure of the input voltage is not apparent in the received current. Sending a telegraph signal via Morse code certainly involved a lot of fine structure, however, and so we naturally wonder: *can* we send a telegraph signal over the Atlantic cable that can be correctly read at the distant receiving end? This is where Thomson's galvanometer really proved its worth, because not only is it an *energy* detector (is there a signal present or not?), it is also a *phase* detector (is the signal positive or negative?). To indicate that a signal is present, the reflected light beam (note 18 in Chapter 5) would deflect from its zero point on the wall; and to indicate the polarity of

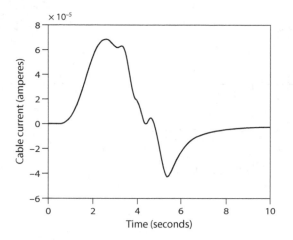

FIGURE 6.2.4. The cable current with (6.2.1) as the input.

the signal, the deflection would be to the right or the left of the zero point.

To send a coded message with a signal that took advantage of this feature of Thomson's galvanometer, the usual Morse code telegraph key was modified: while the original Morse code used dots and dashes in which a dash was simply the key being closed longer than it was for a dot, the Atlantic cable used a key in which a dot was one polarity and a dash was the other polarity (the durations of the dot and the dash were no longer important). We can "experimentally" study how a complicated signal would travel on the Atlantic cable using **fancyinput.m**, under various coding schemes. Just to make up such a scheme, as an example of this, suppose a dot is $+1$ volt and a dash is -1 volt. Each letter (A to Z), each number (0 to 9), and a few additional punctuation characters (! and #, for example) is some combination of dots/dashes (± 1s). To send a character, we send the appropriate ± 1 combination using the following format: each ± 1 is 1 second in duration, and each ± 1 is separated by a 0 (the key connects the sending-end of the cable to ground) for 1 second. After a character has been sent, the key grounds the sending-end of the cable for 2 seconds, to indicate the end of the character. The next character is then sent. And so on.

For example, Figure 6.2.5 shows the input to the cable for the signal sending the character dot-dash-dash-dot-dash, followed by the character

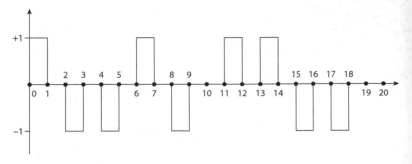

FIGURE 6.2.5. Sending two characters.

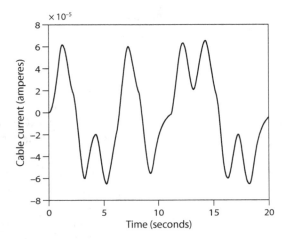

FIGURE 6.2.6. Receiving two characters.

dot-dot-dash-dash (to encode this in **fancyinput.m**, Lines 07 through 13 are replaced with the code in the next shaded box, and *tstop* in Line 01 is set equal to 20). When executed, **fancyinput.m** produced Figure 6.2.6, and I think you can see that the two transmitted characters *could* be correctly read with this coding scheme.

```
for k=1:length(t)
    if t(k)<1
        input(k)=1;
    elseif t(k)<2
        input(k)=0;
    elseif t(k)<3
        input(k)=-1;
    elseif t(k)<4
        input(k)=0;
    elseif t(k)<5
        input(k)=-1;
    elseif t(k)<6
        input(k)=0;
    elseif t(k)<7
        input(k)=1;
    elseif t(k)<8
        input(k)=0;
    elseif t(k)<9
        input(k)=-1;
    elseif t(k)<11
        input(k)=0;
    elseif t(k)<12
        input(k)=1;
    elseif t(k)<13
        input(k)=0;
    elseif t(k)<14
        input(k)=1;
    elseif t(k)<15
        input(k)=0;
    elseif t(k)<16
        input(k)=-1;
    elseif t(k)<17
        input(k)=0;
    elseif t(k)<18
        input(k)=-1;
    else
        input(k)=0;
    end
end
```

Thomson's galvanometer so inspired the imagination of his friend James Clerk Maxwell (note 11 in Chapter 5) that the great Maxwell wrote a poem (!) about it, published anonymously in *Nature*,[14] with the following first two stanzas:

A LECTURE ON THOMSON'S GALVANOMETER

(*Delivered to a Single Pupil in an Alcove with Drawn Curtains.*)

The lamp-light falls on blackened walls,
And streams through narrow perforations;
The long beam trails o'er pastboard scales,
With slow decaying oscillations.
Flow, current! flow! set the quick light-spot flying!
Flow, current! answer, light-spot! flashing, quivering, dying.

O look! how queer! how thin and clear,
And thinner, clearer, sharper growing.
This gliding fire, with central wire
The fine degrees distinctly showing.
Swing, magnet! swing! advancing and receding;
Swing, magnet! answer, dearest, what's your final reading?

Now, to end this book, let me tell you how the signal rate asymmetry of the 1866 Atlantic cable was finally explained by Oliver Heaviside.[15] (Recall that Queen Victoria's celebratory message on the cable took much longer to send than did the American President's longer reply, a puzzle mentioned at the end of Chapter 5.) Just 2 years later, Heaviside personally observed this phenomenon when working in Denmark (in 1868) as a telegraph operator on the then-new 347-mile-long Anglo-Danish cable. In his first year on the job, for example, the transmission speed from England to Denmark was 40% faster than it was in the opposite direction. Similar speed differences were observed on cables linking London and Amsterdam, London and Dublin, and London and Belfast.

To explain these anomalies, Heaviside began by modeling the entire telegraph system with the circuit shown in Figure 6.2.7. As drawn, the

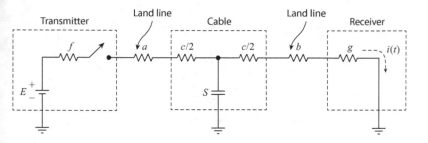

FIGURE 6.2.7. Heaviside's circuit for telegraph transmission from left to right.

telegraph key (at the left) generates pulses to travel to the right (the key is normally pulled up by a spring, and the operator has to push the key down to temporarily close the circuit to a land line (with resistance a ohms). The transmitter is powered by a battery of E volts, and there is a resistance of f ohms associated with the transmitter. The telegraph key connects the transmitter to (as mentioned) a land line with resistance a ohms, which then connects to the cable. The cable itself is modeled as having a total resistance of c ohms, with a capacitance of S farads to ground located (for symmetry) at the cable's midpoint. The right end of the cable connects to a land line with a resistance of b ohms, and then that land line connects to a receiver with an associated resistance of g ohms. The received current is $i(t)$. For transmission in the opposite direction (pulses traveling to the left), we'll simply swap the transmitter (E and f) with the receiver (g).

Heaviside calculated $i(t)$ as follows, with the circuit in Figure 6.2.7 replaced with its (simpler) equivalent shown in Figure 6.2.8. The resistances K_1 and K_2 are given by

$$(6.2.10) \qquad K_1 = \frac{c}{2} + a + f, \ K_2 = \frac{c}{2} + b + g.$$

Applying Kirchhoff's current law at the node at the top of S (marked at voltage v), we have

$$\frac{E-v}{K_1} = S\frac{dv}{dt} + \frac{v}{K_2},$$

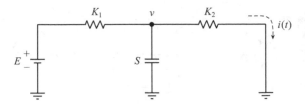

FIGURE 6.2.8. Simplified, equivalent version of Figure 6.2.7.

where the left-hand side is the current flowing into the node from the left, while on the right-hand side are the currents flowing out of the node (through the capacitor to ground, and into the receiver). With a little rearranging, this becomes

$$(6.2.11) \qquad K_1 S \frac{dv}{dt} + \left(1 + \frac{K_1}{K_2}\right) v = E.$$

From the elementary theory of ordinary differential equations, we know the general solution to (6.2.11) is given by a particular solution when the constant voltage E is on the right-hand side, added to the solution to the homogeneous equation (when the right-hand side is set equal to zero). For the particular solution, we argue that a constant voltage E will produce a constant v (with a value we'll call V). Since the derivative of a constant is zero, then (6.2.11) immediately tells us that

$$(6.2.12) \qquad V = \frac{E}{1 + \dfrac{K_1}{K_2}}.$$

For the homogeneous case, we make the usual assumption of an exponential solution, that is, we assume

$$(6.2.13) \qquad v(t) = Ae^{pt},$$

where A and p are some constants. (You'll recall we did this earlier in the book, when solving (3.3.3) for the steady-state temperature of a hot, radiating wire.) If we substitute this $v(t)$ into (6.2.11) with the right-hand side set equal to zero, we get (and I'll let you confirm this)

$$p = -\frac{1}{S\dfrac{K_1 K_2}{K_1 + K_2}}.$$

Thus, the general solution to (6.2.11) is the sum of (6.2.12) and (6.2.13):

$$v(t) = \frac{E}{1 + \dfrac{K_1}{K_2}} + A e^{-\dfrac{t}{S\dfrac{K_1 K_2}{K_1 + K_2}}},$$

and so the received current is

$$i(t) = \frac{v(t)}{K_2} = \frac{1}{K_2} \left[\frac{E}{1 + \dfrac{K_1}{K_2}} + A e^{-\dfrac{t}{S\dfrac{K_1 K_2}{K_1 + K_2}}} \right].$$

Since $i(0) = 0$, we have

$$A = -\frac{E}{1 + \dfrac{K_1}{K_2}},$$

or

$$i(t) = \frac{E}{K_2 + K_1} \left[1 - e^{-\dfrac{t}{S\dfrac{K_1 K_2}{K_1 + K_2}}} \right].$$

This is more compactly written as

$$i(t) = \frac{E}{R} \left[1 - e^{-\frac{t}{T}} \right],$$

where, using (6.2.10), we have

(6.2.14) $$R = K_2 + K_1 = a + b + c + f + g$$

and

(6.2.15) $$T = S \frac{K_1 K_2}{K_1 + K_2}.$$

Heaviside called T the *retardation* (the modern term is *time constant*[16]) of the circuit, because the larger T is, the longer it takes $i(t)$ to rise from its initial value of zero to any given value. The value of T, therefore, plays an obvious role in determining the maximum rate of transmission.

Now, more precisely, the T in (6.2.15) is the T for *left-to-right* transmission, as it was Heaviside's crucial observation that for right-to-left transmission, the value of T *is* different! That is, we should be careful to write (6.2.15) as

(6.2.16) $$T_{l \to r} = S \frac{\left(\dfrac{c}{2} + a + f \right)\left(\dfrac{c}{2} + b + g \right)}{R}.$$

The reason we have to be specific that this is T for the left $\to$ right case is because, for the $r \to l$ case, we swap the locations of the transmitter and receiver. If you look back at Figure 6.2.7 and do that, you'll see that the resulting new version of Figure 6.2.8 looks just like the original Figure 6.2.8 but with the resistances K_1 and K_2 altered to

$$K_1 = \frac{c}{2} + a + g, \ K_2 = \frac{c}{2} + b + f.$$

Thus,

(6.2.17) $$T_{r \to l} = S \frac{\left(\dfrac{c}{2} + a + g \right)\left(\dfrac{c}{2} + b + f \right)}{R}.$$

A comparison of (6.2.16) and (6.2.17) provides a number of insights. For example, if $a = b$ (that is, if the two land lines have equal resistances) then $T_{l \to r} = T_{r \to l}$ no matter what f and g are. Similarly, if $f = g$ (that is, the transmitter and receiver have equal resistances) then

$T_{l \to r} = T_{r \to l}$ no matter what a and b are. It is very unlikely, however, in a real telegraph system, that either of these equalities would hold. In fact, in his 1877 paper, Heaviside gave the following values for the Anglo-Danish cable: on the English side, the land-line resistance was $a = 240$ ohms, while on the Danish side the land-line resistance was $b = 1,250$ ohms. The values for the cable itself were $c = 2,500$ ohms and $S = 120$ microfarads. The value for the transmitter resistance was $f = 150$ ohms, and the receiver resistance was $g = 750$ ohms (contrary to my general assertion, Heaviside stated that for the Anglo-Danish cable, the transmitters and receivers were of identical design in England and Denmark).

What does Heaviside's analysis predict for the difference in the retardation as a function of direction? If we let *left* correspond to England (E) and *right* correspond to Denmark (D), then (6.2.16) and (6.2.17) say

$$(6.2.18) \quad \frac{T_{E \to D}}{T_{D \to E}} = \frac{\left(\dfrac{c}{2} + a + f \right)\left(\dfrac{c}{2} + b + g \right)}{\left(\dfrac{c}{2} + a + g \right)\left(\dfrac{c}{2} + b + f \right)} = \frac{(1,640)\,(3,250)}{(2,240)\,(2,650)} = 0.898.$$

This ratio of the retardations depends only on the resistance values of the receiver, transmitter, cable, and land lines, with *no* dependency on S. The actual, individual values of the retardations, however, do depend on the cable capacitance. In any case, just as observed, there is "less retardation" (that is, faster signaling) in the England-to-Denmark direction compared to the reverse direction. Heaviside's analysis does not predict the observed 40% increase in the transmission speed from England to Denmark, in comparison to the opposite direction, but that failure is almost certainly due to the crude, first-order approximation of Figure 6.2.7 to reality. A first-order explanation for an increase in the transmission speed (in the correct direction) was the central achievement of Heaviside's calculations.

There was a great debate during the 19th century, particularly in England, about the relative strengths that the camp of "theoretical men" versus the camp of "practical men" could bring to the study of complicated electrical problems. Heaviside's 1877 paper was a powerful example of the illumination that could be realized (to quote Heaviside) "when the light of theory is thrown upon" perplexing physical phenomena. If still alive in 1877, Fourier himself would surely have stood and applauded, long and loudly, in enthusiastic agreement with that sentiment.

That is as positive a note as I can imagine on which to end a book such as this one. So, as Porky Pig used to say at the end of many of the 1940/1950s *Looney Toon* cartoons I watched at Saturday afternoon movie theater matinees (where I perhaps spent rather more time than I should have), "Th-th-th-That's all, Folks!"

APPENDIX

How to Differentiate an Integral

Suppose we have a function $g(y)$ defined by the integral

$$g(y) = \int_{v(y)}^{u(y)} f(x,y)\,dx,$$

where, as indicated, the limits on the integral are themselves functions of y, as is the integrand (x, the other variable in the integrand, is a dummy variable of integration). What's the derivative of $g(y)$ with respect to y? This is a generalization of a question that arises in Chapter 5, and again in Chapter 6, when the calculation of the current in the Atlantic cable makes appearances: see (5.4.3) and (6.2.8). The answer to our question is usually first taken up in an advanced calculus course but, in fact, if we are willing to make some plausible assumptions, then it can be done using just the elementary ideas developed in high school AP-calculus.

From the definition of the derivative, we have

(A1)
$$\frac{dg}{dy} = \lim_{\Delta y \to 0} \frac{g(y+\Delta y) - g(y)}{\Delta y}.$$

Since the integration limits depend on y, then a Δy will generally cause there to be a Δv and a Δu. So, the numerator of (A1) can be written as

$$g(y+\Delta y) - g(y) = \int_{v+\Delta v}^{u+\Delta u} f(x,y+\Delta y)\,dx - \int_{v}^{u} f(x,y)\,dx$$

$$= \left[\left\{ \int_{v+\Delta v}^{v} + \int_{v}^{u} + \int_{u}^{u+\Delta u} \right\} f(x,y+\Delta y)\,dx \right] - \int_{v}^{u} f(x,y)\,dx,$$

or

(A2)
$$g(y+\Delta y) - g(y) = \int_v^u \{f(x,y+\Delta y) - f(x,y)\}\,dx$$
$$+ \int_u^{u+\Delta u} f(x,y+\Delta y)\,dx$$
$$- \int_v^{v+\Delta v} f(x,y+\Delta y)\,dx,$$

where the minus sign occurs in front of the last integral because (note carefully!) we've reversed the integration limits.

Now, as $\Delta y \to 0$, we expect that $\Delta u \to 0$ and $\Delta v \to 0$, too. Thus, in the limit $\Delta y \to 0$, we see that (A2) becomes

$$\lim_{\Delta y \to 0} \{g(y+\Delta y) - g(y)\} = \lim_{\Delta y \to 0} \int_v^u \{f(x,y+\Delta y) - f(x,y)\}\,dx$$
$$+ f(u,y)\Delta u - f(v,y)\Delta v,$$

where the last two terms follow because, as $\Delta v \to 0$ and $\Delta u \to 0$, the value of x over the entire integration intervals of the last two integrals in (A2) remains essentially unchanged at $x=v$ or at $x=u$, respectively. Thus, (A1) becomes

(A3)
$$\frac{dg}{dy} = \lim_{\Delta y \to 0} \frac{g(y+\Delta y) - g(y)}{\Delta y}$$
$$= \lim_{\Delta y \to 0} \frac{1}{\Delta y} \int_v^u \{f(x,y+\Delta y) - f(x,y)\}\,dx$$
$$+ \lim_{\Delta y \to 0} f(u,y)\frac{\Delta u}{\Delta y} - \lim_{\Delta y \to 0} f(v,y)\frac{\Delta v}{\Delta y}.$$

Finally, taking the $\frac{1}{\Delta y}$ inside the integral,[1] in the limit, (A3) becomes

$$\frac{dg}{dy} = \int_v^u \lim_{\Delta y \to 0} \frac{f(x,y+\Delta y) - f(x,y)}{\Delta y}\,dx + f(u,y)\frac{du}{dy} - f(v,y)\frac{dv}{dy},$$

or

$$\text{(A4)} \qquad \frac{dg}{dy} = \int_v^u \frac{\partial f}{\partial y} dx + f(u,y)\frac{du}{dy} - f(v,y)\frac{dv}{dy},$$

a result called *Leibniz's formula*,[2] after the German mathematician Gottfried Wilhelm Leibniz who, you'll recall, appeared in the opening section of Chapter 1 as the discoverer of a beautiful infinite series expression for $\frac{\pi}{4}$. (A4) shows that if the integration limits are not functions of the independent variable (y), then the derivative of the integral is simply the integral of the derivative. If the limits are functions of y, however, additional terms come into play.

As an example of the utility of Leibniz's formula, let's use it to derive the result that was simply given without proof in note 11 of Chapter 4, when discussing the definition of the error function:

$$\int_0^\infty e^{-y^2} \, dy = \frac{\sqrt{\pi}}{2}.$$

We start by defining the function[3]

$$g(t) = \left\{ \int_0^t e^{-y^2} \, dy \right\}^2,$$

and so what we are after is

$$\sqrt{g(\infty)} = \int_0^\infty e^{-y^2} \, dy.$$

Differentiating $g(t)$ using Leibniz's formula (notice that the upper limit on the $g(t)$ integral is a function of t, while the integrand and the lower limit are not such functions), we get

$$\frac{dg}{dt} = \frac{d}{dt}\left\{ \int_0^t e^{-y^2} \, dy \right\}^2 = 2\left\{ \int_0^t e^{-y^2} \, dy \right\}\left\{ \frac{d}{dt}\int_0^t e^{-y^2} \, dy \right\} = 2\left\{ \int_0^t e^{-y^2} \, dy \right\}\{e^{-t^2}\},$$

because

$$\frac{\partial e^{-y^2}}{\partial t} = 0.$$

Now, taking the e^{-t^2} factor inside the integral (which we can do as y, not t, is the integration variable),

$$\frac{dg}{dt} = 2\int_0^t e^{-(t^2+y^2)} \, dy.$$

Next, change the variable to

$$x = \frac{y}{t},$$

which means that, as y varies from 0 to t, x varies from 0 to 1. Thus, as $y = xt$ ($dy = t \, dx$), we have

$$\frac{dg}{dt} = 2\int_0^1 e^{-(t^2+x^2t^2)} t \, dx,$$

or

$$\frac{dg}{dt} = \int_0^1 2te^{-(1+x^2)t^2} \, dx.$$

Notice that the integrand of this integral can be written as a partial derivative:

$$2te^{-(1+x^2)t^2} = \frac{\partial}{\partial t}\left\{-\frac{e^{-(1+x^2)t^2}}{1+x^2}\right\}.$$

Thus,

$$\frac{dg}{dt} = \int_0^1 \frac{\partial}{\partial t}\left\{-\frac{e^{-(1+x^2)t^2}}{1+x^2}\right\} dx = -\frac{d}{dt}\int_0^1 \frac{e^{-(1+x^2)t^2}}{1+x^2} dx.$$

This last expression can be integrated by inspection to give

$$g(t) = -\int_0^1 \frac{e^{-(1+x^2)t^2}}{1+x^2}\,dx + C,$$

where C is the constant of integration. We can find C as follows. Setting $t=0$ in our original definition of $g(t)$, we have

$$g(0) = \left\{ \int_0^0 e^{-y^2}\,dy \right\}^2 = 0,$$

while, if we remember (1.3.3), our last result says

$$g(0) = -\int_0^1 \frac{1}{1+x^2}\,dx + C = -\tan^{-1}(x)\Big|_0^1 + C = -\{\tan^{-1}(1) - \tan^{-1}(0)\} + C$$

$$= -\frac{\pi}{4} + C.$$

So, setting our two expressions for $g(0)$ equal, we have $C = \frac{\pi}{4}$ and therefore,

$$g(t) = -\int_0^1 \frac{e^{-(1+x^2)t^2}}{1+x^2}\,dx + \frac{\pi}{4}.$$

Now, let $t \to \infty$. The integral clearly vanishes over the entire interval of integration, and so

$$g(\infty) = \frac{\pi}{4}.$$

The object of our pursuit is, you'll recall, $\sqrt{g(\infty)}$, and we have

$$\sqrt{g(\infty)} = \int_0^\infty e^{-y^2}\,dy = \sqrt{\frac{\pi}{4}} = \frac{\sqrt{\pi}}{2}.$$

As a second example, this time with an integral in which both limits are constants, consider the evaluation of

$$\int_0^1 t^n \ln(t)\,dt, \quad n > -1.$$

We start by defining the integral

$$I(n) = \int_0^1 t^n\,dt = \left(\frac{t^{n+1}}{n+1}\right)\Bigg|_0^1 = \frac{1}{n+1}, \quad n > -1.$$

The reason this is useful is because we can write $I(n)$ as

$$I(n) = \int_0^1 e^{\ln(t^n)}\,dt = \int_0^1 e^{n\ln(t)}\,dt,$$

and so, differentiating under the integral sign with respect to n,

$$\frac{dI}{dn} = \int_0^1 \ln(t)e^{n\ln(t)}\,dt = \int_0^1 t^n \ln(t)\,dt.$$

But since

$$\frac{dI}{dn} = -\frac{1}{(n+1)^2},$$

then, just like that (and so fast you perhaps can hardly believe we are done),

$$\int_0^1 t^n \ln(t)\,dt = -\frac{1}{(n+1)^2}, \quad n > -1.$$

This result is equally easy to establish with an integration by parts.

For a third (and final) example of differentiating an integral, consider the problem of evaluating

$$\int_0^\infty \frac{\cos(x)}{(x^2+b^2)^2}\,dx.$$

This integral appeared some years ago in a text[4] as an example of the power of Cauchy's *contour integration*, a method that is definitely *not* AP-calculus! (See, too, the discussion in this book of Cauchy's integral, just after Figure 2.3.2.) What I'll show you now is a way to do this integral that is within the reach of a freshman calculus student if we allow the use of Leibniz's formula. In fact, what we'll evaluate is the slightly more general integral

$$\int_{-\infty}^\infty \frac{\cos(ax)}{(x^2+b^2)^2}\,dx,$$

and so the particular case of $a=1$ will be the original integral.

We start with a result due to Fourier's colleague in the French Academy, Laplace, dating from 1810:

$$\int_0^\infty \frac{\cos(ax)}{x^2+b^2}\,dx = \frac{\pi}{2b}e^{-ab}.$$

Because the integrand is even, this says[5]

$$\int_{-\infty}^\infty \frac{\cos(ax)}{x^2+b^2}\,dx = \frac{\pi}{b}e^{-ab}.$$

Thus, differentiating with respect to the parameter b, we have

$$\int_{-\infty}^\infty \frac{-\cos(ax)2b}{(x^2+b^2)^2}\,dx = \pi\left[\frac{-abe^{-ab}-e^{-ab}}{b^2}\right],$$

or

$$\int_{-\infty}^\infty \frac{\cos(ax)}{(x^2+b^2)^2}\,dx = \pi\left[\frac{abe^{-ab}+e^{-ab}}{2b^3}\right],$$

and so

$$\int_{-\infty}^{\infty} \frac{\cos(ax)}{(x^2+b^2)^2}\,dx = \frac{\pi(1+ab)}{2b^3 e^{ab}}.$$

We could, of course, continue to differentiate with respect to either a or b until the cows come home, and thereby derive an endless sequence of integral formulas. I'll leave doing *that* for your amusement!

ACKNOWLEDGMENTS

Writing is a lonely game. By that I mean writing a lengthy, technical (that is, *mathematical*) discussion is not generally a good arena for collaboration. Now, some people *can* pull it off—the famous joint efforts of the English mathematicians G. H. Hardy (1877–1947) and J. E. Littlewood (1885–1977) being a classic example. But even Littlewood had some qualifying feelings on the experience of collaboration, as illustrated by a funny story he once told about some joint work he did during the Second World War: "Two rats fell into a can of milk. After swimming for a time one of them realized his hopeless fate and drowned. The other persisted and at last the milk was turned to butter and he could get out." (Littlewood doesn't say which rat he was, but I suspect he was the expired, milk-logged rodent last seen floating belly-up.)[1]

Jumping forward into modern times, as a literary critic recently wrote in a hilarious evaluation of a multi-author novel, "Writing, like dying, is one of those things that should be done alone or not at all. In each case, loved ones may hover around and tender their support, but, in the end, it's up to you."[2] Unlike dying, however, being the lone author of a technical math/physics book, while relentlessly demanding, can also be an enormously pleasant act of creation. I've had lots of fun writing *all* of my books, but this one is the best of all (for me, anyway—for my readers, I can only hope).

But of course I didn't do absolutely *everything* by myself. There were, for example, all the folks at Princeton University Press, who had a lot more than "just something" to do with this book being in your hands. My deepest and appreciative thanks, therefore, go to my editors Susannah Shoemaker and Vickie Kearn (and their hard-working assistant Lauren Bucca), to production editor Debbie Tegarden (with whom I've spent literally hours on the telephone over the many years we've worked together, discussing book art and book writing in general), to artist Laurel Muller, and to copyeditor Cyd Westmoreland. I am also most grateful to two anonymous reviewers who pointed out a number of ways to improve my original efforts, and I followed their excellent advice without exception. Then, four more anonymous reviewers gave my "final" submission yet more careful scrutiny. Judith Grabiner, professor emerita of mathematics at Pitzer College, Claremont, California (across the street— almost—from Harvey Mudd College where, decades ago, I once taught), graciously agreed to contribute a Foreword to the book, written in part to honor the memory of our mutual mathematical friend, the late Barbara Beechler.

To the people at The MathWorks in Natick, MA, the creators of *MATLAB* and the Symbolic Math Toolbox, who gave me a free license to use that software to produce all of the computer plots in this book, I owe many thanks. And all the tiring hours once spent on previous books hunting through musty library stacks for old scientific journals (followed by even more hours bent over hot, balky photocopying machines) have been replaced by the vast, on-line digital database archives maintained by the hardworking library staff of the University of New Hampshire's Dimond Library. Now, with just the touch of a finger on a keyboard in the comfort of my home office, I can sip a cup of tea and munch a croissant, all the while my computer's printer produces an immaculate copy of just about any scientific paper ever published.

The entire book was written in that most excellent of coffee shops, *Me & Ollie's Bread, Bakery and Café* on Water Street in Exeter, New Hampshire. *Ollie's* ever-pleasant staff kept my coffee mug in an endless state of "floweth-over," with the best hot coffee anywhere. With retrosongs by Buddy Holly, Buddy Knox, Chuck Berry, Marty Robbins, and Patsy Cline playing over the sound system, and a constant traffic of high school kids from Phillips Exeter Academy from just up the street, it felt just like I was back in my own high school days of the mid-1950s.

All of the above were essential to the physical reality of this book. But none of that would have been enough, alone, without the support and loving companionship of my wife of nearly 60 years, Patricia Ann. Without Pat, I might well have ended up as a video gamer guy, wearing a virtual reality helmet and blasting away with a grenade launcher at armies of crazed, dead-alive zombies on a TV screen. Instead, because of Pat, I am a reasonably fit fellow, sitting in an ergonomic chair in a well-lit room typing partial differential equations on a computer screen. Now that might seem like a toss-up choice to some but, believe me, there *is* a significant difference!

Paul J. Nahin
Exeter, New Hampshire
April 2019

NOTES

Chapter 1

1. Quote from Peter Lax's 2007 Gibbs Lecture to the American Mathematical Society, titled "Mathematics and Physics." See the *Bulletin of the American Mathematical Society*, January 2008, pp. 135–152.
2. The friend was the Hungarian mathematician Marcel Grossman (1878–1936), who let Einstein (a casual sort of fellow who was known to skip a class here and there, now and then) study his meticulous notes when both were students at Zurich Polytechnic in Switzerland. Grossman later became head of the math department at Zurich Polytechnic.
3. Quoted from Walter Isaacson, *Einstein, His Life and Universe*, Simon & Schuster 2007, p. 193.
4. Fourier was known at the highest level in French politics, having served as a scientific advisor to Napoleon himself, when France invaded Egypt in 1798. Fourier returned to France in 1801, and may have begun pondering the physics of heat soon after.
5. Fourier's intense interest in heat was not without a touch of irony. For most of his adult life, he was apparently bothered by rheumatism and tried to self-treat it by wearing numerous layers of clothing to keep warm. He is quoted in Francois Arago's *Biographies of Distinguished Scientific Men* (1857) as saying "One would suppose me to be corpulent [but be assured] that there is much to deduct from this opinion. If after the example of the Egyptian mummies, I was subjected to the operation of disembowelment—from which heaven preserve me—the residue would be found to be a very slender body." As a final note on Fourier the man, he didn't die from the lack of warmth, but rather of a heart attack. Arago records his last words as being far less memorable than were Einstein's: "Quick, quick! Some vinegar. I am fainting!"
6. You can find an extended discussion of the Laplace transform in my book *Transients for Electrical Engineers*, Springer 2018, along with computer codes to do all the (often vastly) grubby arithmetic. The use of the transform in solving Atlantic cable questions receives a lot of attention in that book.
7. See the shaded box in the text for a freshman calculus derivation of Euler's identity, and my books *An Imaginary Tale: The Story of $\sqrt{-1}$*, Princeton 2016, and *Dr. Euler's Fabulous Formula*, Princeton 2011, for lots more on this wonderful identity.
8. Electrical engineers call Figure 1.3.3 a *square wave*, and the reason for that is surely obvious from the plot.

9. You can find an extended discussion of both Wilbraham the man, and of his remarkable discovery of the Gibbs phenomenon 50 years before Gibbs came on the scene, in *Dr. Euler* (note 7), pp. 163–173. We'll encounter the Gibbs phenomenon again in the next chapter.

10. Integrate the series for $\frac{\pi}{4}x$ in (1.3.10), and see if you can use the result to conclude that $1 - \frac{1}{3^3} + \frac{1}{5^3} - \frac{1}{7^3} + \cdots = \frac{\pi^3}{32}$. Alas, this calculation does *not* allow the further evaluation of $\frac{1}{1^3} + \frac{1}{2^3} + \frac{1}{3^3} + \frac{1}{4^3} + \cdots$, one of today's most famous unsolved problems in mathematics. It utterly defeated Euler, and everybody else since.

11. A more recent example of a surprising appearance of pi-squared is the following theorem, due to the Swedish mathematician Fritz David Carlson (1888–1952), who in 1935, showed that if $a_n > 0$ and if all the series exist, then

$$\left\{\sum_{n=1}^{\infty} a_n\right\}^4 < \pi^2 \left\{\sum_{n=1}^{\infty} a_n^2\right\} \left\{\sum_{n=1}^{\infty} n^2 a_n^2\right\}.$$ To pursue this would take us too far afield from Fourier and heat, and Thomson and electric telegraph cables, but perhaps it's worth a final comment to tell you that Carlson's inequality can be established with just a few lines of AP-calculus, with the π^2 showing up from seemingly out of nowhere.

12. After the French-born mathematician Abraham De Moivre (1667–1754), who was a friend of the great English mathematical physicist Isaac Newton (1642–1727). In a 1698 paper published in the *Philosophical Transactions* of the Royal Society, De Moivre mentions that Newton knew an equivalent to (1.3.14) as early as 1676.

13. *An Imaginary Tale* (note 7), pp. 56–60.

14. If x and y are both real, the *conjugate* of the complex quantity $x + iy$, written as $(x + iy)^*$, is $x - iy$. That is, we simply reverse the sign of the imaginary part.

15. After the French mathematician Siméon-Denis Poisson (1781–1840). The word *kernel* is used because the quantity occurs as the *core* portion of the integrand of an important integral due to Poisson.

Chapter 2

1. This is not always true, but it *is* true for J, because neither of the integration limits is a function of a_n. For the general case, to differentiate an integral we would use *Leibniz's formula* (see the Appendix), but we don't need it here (we will use it in Chapters 5 and 6). There are also details about what mathematicians call *uniform convergence* concerning the J integral, which we'll ignore (remember my earlier words about assumptions).

2. It's a curious fact that, while the conditions of evenness and oddness are quite restrictive, *any* function can always be written as the sum of an even function and an odd function. The proof is easy, direct, and convincing, as it's a proof by construction (the best kind of all). First, whatever $f(t)$ is, $f(t) + f(-t)$ is even, and $f(t) - f(-t)$ is odd. Then simply observe that $f(t) = \frac{1}{2}[f(t) + f(-t)] + \frac{1}{2}[f(t) - f(-t)]$. Done!

3. See note 14 in Chapter 1.

4. These two claims are easy to establish, and you should do that.

5. After the French mathematician Marc-Antoine Parseval des Chenes (1755–1836), who published his formula (in much different form, in a non-Fourier series context) in 1799, when Fourier was still in Egypt with Napoleon.

6. As a hint to get you started, make a power series expansion of $e^{-2\pi p}$, and retain the terms up to the one in p^3. From that point on, it's just a matter of being careful with the algebra.

7. The rule is named for the French writer G.F.A. de L'Hôpital (1661–1704), who included it in the first-ever textbook (1696) on the differential calculus. The rule is, however, actually due to the Swiss mathematician Jean (John) Bernoulli (1667–1748), who will appear in Chapter 4 with the inspiration for the key idea behind solving the heat equation. The rule says that if $f(x)$ and $g(x)$ are differentiable functions that both vanish at $x=a$, then the indeterminate $\lim\limits_{x \to a} \dfrac{f(x)}{g(x)} = \dfrac{0}{0}$ can be calculated as $\lim\limits_{x \to a} \dfrac{df/dx}{dg/dx}$.

8. The vanishing at $x=\pi$ is obvious, while at $x=0$, we get an indeterminate result of $\infty \cdot 0$ (but L'Hôpital's rule again comes to the rescue and shows that the value at $x=0$ is zero).

9. In two papers, published in 1909 and 1916, the great English mathematician G. H. Hardy (1877–1947) wrote at length on a number of different ways to show that $\int_0^\infty \dfrac{\sin(\lambda)}{\lambda} d\lambda = \dfrac{\pi}{2}$. Those papers are reprinted in *The G. H. Hardy Reader*, Cambridge University Press 2015, pp. 311–321. In the next shaded box, you'll find yet a different derivation, one not in Hardy's discussions. My feeling is that Hardy would have found much to not like in the box derivation, but that's the difference between a pure mathematician and an engineer.

10. After the German mathematician Gustav Dirichlet (1805–1859).

11. See my *Inside Interesting Integrals*, Springer 2015, pp. 231–234 for details.

Chapter 3

1. After the Austrian mathematical physicists Joseph Stefan (1835–1893), and his student Ludwig Boltzmann (1844–1906), who articulated it in 1879–1884.

2. However, notice that $u^4 - u_0^4 = (u^2 - u_0^2)(u^2 + u_0^2) = (u - u_0)(u + u_0)(u^2 + u_0^2)$, and so, if $u \approx u_0$, then $u + u_0 \approx 2u_0$ and $u^2 + u_0^2 \approx 2u_0^2$. Thus, if $u \approx u_0$ then $u^4 - u_0^4 \approx 4u_0^3(u - u_0)$, which is *linear* in the temperature difference. This is called *Newton's law of cooling* (after you-know-who), and it leads to all the problems you'll find in textbooks, because they *can* be easily solved (see, for example, the final section of this chapter).

3. In the metric cgs (centimeter-gram-second) system, k has the units of cm^2/s. The cgs value of k varies from 0.001 (rubber, for example) to almost 2 (metals) for the various forms of matter commonly encountered on Earth.

4. Thomson was not the originator of the idea of estimating the age of the Earth from a study of cooling spheres. A century before, the French naturalist Georges-Louis

Leclerc, Comte de Buffon (1707–1788), a man well known to students of probability for his famous demonstration of how to *experimentally* determine the value of pi by randomly tossing a needle on a regularly lined surface, had performed an equally dramatic experiment concerning cooling iron spheres. Using his bare hands (!) as a "thermometer" to estimate cooling rates of initially incandescent (!!) globes, only inches in diameter, to that of a globe the size of the Earth, Buffon arrived at an age of 75,000 years.

5. Harold C. Urey, *The Planets: Their Origin and Development*, Yale University Press 1952, p. 163. Urey (1893–1981) was the 1934 recipient of the Nobel prize in chemistry.

6. Darwin expressed his concerns in 1869 as follows: "I am greatly troubled at the short duration of the world according to Sir W. Thomson, for I require for my theoretical views a very long period *before* the Cambrian formation [when Earth's biology blossomed]." See William Thomson, "On the Secular Cooling of the Earth," reprinted in his *Mathematical and Physical Papers* (volume 3), Cambridge University Press, 1890, pp. 295–311.

7. For a scholarly treatment of the debate, see Joe D. Burchfield, *Lord Kelvin and the Age of the Earth*, University of Chicago Press, 1975 (a new edition appeared in 1990).

8. George F. Becker, "Age of a Cooling Globe in Which the Initial Temperature Increases Directly as the Distance from the Surface," *Science*, February 7, 1908, pp. 227–233.

9. You can find an extended discussion on how Thomson used gravitational contraction to explain the energy of the Sun in my book, *Mrs. Perkins's Electric Quilt*, Princeton University Press, 2009, pp. 157–162. A detailed description of the thermonuclear fusion reactions that are the actual source of stellar energy (put forth in the 1930s) won the 1967 Nobel physics prize for the German-American physicist Hans Bethe (1906–2005).

10. These words appear on page 90 of Alexander Freeman's translation of *Analytical Theory*. The italicized words are Fourier's way of ensuring that the air in contact with the sphere's surface doesn't heat up, but rather is always at temperature 0°C.

11. If the finite mass is in the shape of a circular disk (cylinder) with a given radius, with its two circular faces insulated while the cylindrical boundary surface is kept at temperature 0°C, and the mass has an initial temperature distribution that varies only as a function of the radial distance r from the center of the disk towards the cylindrical boundary surface, then $\frac{\partial u}{\partial t} = k \left[\frac{\partial^2 u}{\partial r^2} + \left(\frac{1}{r} \right) \frac{\partial u}{\partial r} \right]$, which is very similar (but *not exactly identical*) to the heat equation for a sphere. You should now be able to show this by modifying the analysis for a sphere in the obvious way, and I'll leave the details of doing that as a challenge problem.

12. At $x = 2l$ and $x = 3l$, for example, (3.3.9) tells us that, for $U_0 = 100°C$ (and beeswax), the wire's steady-state temperatures are 39.7°C and 25°C, respectively.

Chapter 4

1. You might be wondering how Bernoulli came up with this trick in 1753, a half-century before Fourier developed the heat equation. Bernoulli was actually studying a different second-order partial differential equation (called the *wave equation*) that describes a vibrating string. The same trick works for the heat equation, too. You can find more discussion on the wave equation and vibrating strings in *Dr. Euler* (note 7, Chapter 1), pp. 114–128. We'll solve the wave equation in Chapter 6, when we get to the topic of telegraph cables.

2. This may sound like something only an engineer would write, but a book by *mathematicians* agrees. As R. L. Borelli and C. S. Coleman state in their *Differential Equations*, Prentice-Hall 1987, p. 508: "It will come as no surprise at all . . . that all the partial differential equations which arise in models of *physical phenomena* [my emphasis] behave 'nicely.'" That is, they have unique solutions for specific boundary/initial conditions.

3. Mathematicians have developed proofs of this uniqueness, which we'll take as being physically obvious. If you are interested in how to prove our "obvious" assumption, then take a look (for example) at R. V. Churchill, *Fourier Series and Boundary Value Problems*, McGraw-Hill 1941, pp. 105–108.

4. Urey's book on planet formation (note 5 in Chapter 3), on its page 53, gives the answer to a more general problem than is solved here. His result assumes that radioactive heat sources are scattered about in the sphere: if all the sources are set to zero, his result reduces to (4.2.9). Urey doesn't provide the details of how his result was obtained, other than to state the Laplace transform was used (see note 6 in Chapter 1). It is historically interesting to note that Urey thanks "S. Chandrasekhar" for help in solving the heat equation in various chapters of his book: Subrahmanyan Chandrasekhar (1910–1995) was an Indian mathematical physicist who, 31 years after Urey's book was published, won the 1983 Nobel prize in physics (for, in part, showing that any star with mass greater than 1.4 times that of the Sun would eventually explode as a supernova, and so inject all the elements essential for life into the universe).

5. The series $1-1+1-\cdots$ is called *Grandi's series*, after the Italian mathematical theologian Guido Grandi (1671–1742) who discussed it in 1703. The averaging idea, for assigning a value to the "sum" of the series, has become accepted mathematics, as either a *Cesàro sum* or an *Abel sum*, after the Italian Ernesto Cesàro (1859–1906) and the Norwegian Niels Abel (1802–1829). Grandi "derived" $1-1+1-\cdots=\frac{1}{2}$ by writing $1-x+x^2-x^3+\cdots=\frac{1}{1+x}$ and then setting $x=1$. Observing that the series can alternatively be written as $(1-1)+(1-1)+\cdots=0+0+\cdots=0$, Grandi argued that this proved the world could be created out of nothing.

6. You should confirm this calculation (it's easy to do).

7. Can you give a physical reason for the infinity? Think about this.

8. This is "physicist humor."

9. This definite integral has been known for a very long time: it appears in Fourier's *Analytical Theory* (see page 373 in Freeman's translation). If you are interested in the details of how such an integral can be evaluated, see my *Inside Interesting Integrals* (note 11 in Chapter 2), pp. 77–79.

10. The error function has been around in mathematics a long time, but it didn't have a formal name until 1871, when the English mathematician James Glaisher (1848–1928) so christened it.

11. $\mathrm{erf}(\infty) = 1$, because $\int_0^\infty e^{-y^2} dy = \frac{\sqrt{\pi}}{2}$, which is derived in the Appendix (using a result called *Leibniz's formula* that will be of great aid in the next two chapters).

12. Notice, carefully, that in Figure 4.4.1, we are measuring x in both the positive and negative directions from a point designated as $x = 0$. So, $-L \leq x \leq L$, where $x = -L$ and $x = L$ are physically the same point.

13. On pages 271–273 of Freeman's Dover Publications translation of *Analytical Theory*.

14. If we call the roots $\frac{\mu_n}{R}$, then the first four values of μ_n are: $\mu_1 = 4.493409$, $\mu_2 = 7.725251$, $\mu_3 = 10.904121$, and $\mu_4 = 14.066193$. Can you calculate μ_5 to six decimal places? In Chapter 6, I'll show you how to do it, using an approach Fourier could have fantasized about.

15. The infinite set of functions $\sin(\lambda_n r)$, $n = 1, 2, 3, \ldots$ form what mathematicians call an *orthogonal* set on the interval $0 \leq r \leq R$, where the λ_n are the solutions to (4.5.6).

Chapter 5

1. Wells was (by far) the more fantastic of the two, never letting the laws of physics get in the way of telling a good story. That apparently greatly irritated Verne, whose signature characteristic as an adventure writer was a faithful adherence to the known laws of science. Verne, in fact, wasn't above actually sprinkling some math here and there in his tales. The modern publishing view, that for every additional equation in a text, the potential audience is cut in half, didn't apply to Verne and, to their credit, his readers *loved* it! (What does this say about modern education?)

2. An undersea telegraph cable was, in 1854, an accomplished fact, with the success of the 1851 cable between Dover, England and Calais, France, already three years in operation. But that cable was only 26 miles long, across the English Channel at a maximum depth that didn't quite reach 600 feet. The Atlantic cable would be over seventy-five times longer and experience depths more than twenty times greater, and so would be proportionally (*at least*, and more likely *exponentially*) more difficult to construct.

3. Or, as I just did as I typed this, yell "Hey, Google, what's the temperature right now in London?" and get the answer from my Google Assistant gadget in less than 2 seconds.

4. That possibility was realized. Between the end of July and the end of October in 1867, for example, the ultimately successful cable of 1866 transmitted 2,772 messages. The charge was $10 a word. With a 10-word minimum, each message cost at least $100, and so the total revenue for those 3 months was in excess of $277,000, or well over more than 1 million dollars per year.

5. This is primarily a book of mathematical physics, and so I have (greatly!) compressed the historical tale of how the Atlantic cable came to be. It's a wonder-

fully romantic story, however, and two particularly good references that I can recommend are Bern Dibner, *The Atlantic Cable*, Burndy Library, 1959, and W. H. Russell, *The Atlantic Telegraph*, Nosuch Publishing, 2005 (originally published in 1866). Dibner was a modern historian, while William Howard Russell (1820–1907) was a contemporary journalist who actually sailed on the *Great Eastern* when she laid the 1866 cable. His detailed accounting of that adventure includes brief exchanges he had with William Thomson, himself, who was also on board.

6. From a speech at Washington University in St. Louis, delivered April 1857 by Edward Everett (1794–1865).

7. Such was Thomson's reputation that, when he died in 1907, he was entombed just steps away from Isaac Newton's grave in Westminster Abbey, a very special resting place reserved for England's greatest heroes. As I write, the famous theoretical physicist Stephen Hawking (1942–2018) has also just received that honor.

8. Sources are *not* passive, as they are the origins of energy in an electrical circuit. A constant voltage source maintains a fixed voltage drop across its terminals, independent of the current in it (think of the common battery). Constant current sources maintain a fixed current in themselves, independent of the voltage drop across their terminals, and are *not* something you can buy in the local drugstore like a battery. You have to construct them. Batteries are the only sources we'll consider in this book.

9. It's easy to write the electronic charge is $e = -1.6 \times 10^{-19}$ coulombs, but what is a *coulomb*? We can't just keep defining electrical units in terms of other electrical units without ending up in a circular trap. At *some* point, we have to tie an electrical quantity to the physical world. One way to do that for the coulomb, for example, is to define it as that charge required to precipitate a particular quantity of silver out of a silver nitrate solution (via electrolysis). You can find videos of this on YouTube.

10. I've taken this story from a book by two mathematicians, Peter G. Doyle and J. Laurie Snell, *Random Walks and Electrical Networks*, Mathematical Association of America, 1984, p. 70.

11. This is, of course, the great Scottish mathematical physicist James Clerk Maxwell (1831–1879), who was one of William Thomson's friends. The famous *Maxwell's equations* are his, about which Richard Feynman (remember Feynman's trick about differentiating an integral from Chapter 2?) wrote, in his famous multivolume undergraduate text *Feynman's Lectures on Physics*, "Ten thousand years from now there can be little doubt that the most significant event of the 19th century will be judged as Maxwell's discovery of the laws of electrodynamics. The American Civil War will pale into provincial insignificance in comparison with this important scientific event of the same decade."

12. You can gain an idea of the confusion in Maxwell's mind about electricity in P. M. Heimann, "Maxwell, Hertz, and the Nature of Electricity," *Isis*, Summer 1971, pp. 149–157. See also my paper, "Oliver Heaviside: An Accidental Time Traveler," *Philosophical Transactions A of the Royal Society*, December 2018.

13. Oppenheimer was the scientific head of the American atomic bomb program in World War II and, despite reservations on the morality of creating such a

horrendous weapon of mass destruction, he and others did so anyway, at least in part because the bomb presented problems that were "technically sweet."

14. Lord Rayleigh made this statement in his famous 1894 book *The Theory of Sound*. Since there are 6,076 feet in a nautical mile, there are 1.85×10^5 cm in a nautical mile, and it is then an easy calculation to show that the values of R and C given in the text work out to give a product of 4.4×10^{-17}, pretty close to Rayleigh's value.

15. William Thomson, "On the Theory of the Electric Telegraph," *Proceedings of the Royal Society of London*, May 1855, pp. 382–399. There is a note (dated July 2, 1852) in one of Thomson's research notebooks that indicates he *did* initially consider induction, but eventually concluded it too unimportant to even mention in his 1855 analysis. See D. W. Jordan, "The Adoption of Self-Induction by Telephony, 1886–1889," *Annals of Science*, September 1982, pp. 433–461.

16. John M. Picker, "The Atlantic Cable," *Victorian Review*, Spring 2008, pp. 34–38.

17. Daniel R. Headrick and Pascal Griset, "Submarine Telegraph Cables: Business and Politics, 1838–1939," *Business History Review*, Autumn 2001, pp. 543–578.

18. The detection device used on the Atlantic cable was the *marine* (or *mirror*) galvanometer. The construction was as a tiny magnet hanging by a silk thread inside a coil of wire carrying the received current. A mirror was attached to the magnet, and the magnet/mirror assembly would be rotated by the torque produced by the interaction of the magnet's field with the magnetic field produced by the received current. The rotation was detected by bouncing a light beam off the mirror, and observing the motion of a reflected spot of light on a wall. This clever gadget was yet another of Thomson's many inventions.

19. In contrast, (5.4.2) shows that the cable voltage does *not* have a peak value in response to a unit step voltage input, but rather simply increases monotonically from zero to 1 volt. Having a peak allows a more definitive observation of a received signal than does a monotonic increase.

Chapter 6

1. You can find details of this story in Claire Tomalin, *The Invisible Woman: The Story of Nelly Ternan and Charles Dickens*, Knopf, 1991 (see, in particular, pp. 179–180 for Dickens' coded message).

2. Such a signal is, in the limit $T \to 0$, called an *impulse*, and it is a routine part of modern physics and electrical engineering. In the 1850s, however, Thomson was far ahead of the times with his theoretical use of such a signal.

3. When dealing with trigonometric functions in differential equations, it is almost always useful to express them in complex exponential form, because exponentials are so easy to differentiate. We do all the analysis with the complex exponentials and then, as the observed physical variables (voltages and currents) are real valued, we keep only the real part of the resulting complex-valued answer.

4. The Atlantic cable was what is now called a *dispersive medium*. A modern example is the ocean itself, and that is the reason it is not possible to communicate via voice radio with submerged submarines at sea. Radio communication through sea water is achieved only by using what are called *extremely low frequencies* (ELF),

typically less than 100 Hz, which results in the transmission of information at low speed (this has important implications for the remote command-and-control of the weapons systems on submerged submarines).

5. I am greatly abbreviating history here, but you can find the detailed (often incredible) tale of Heaviside in my book *Oliver Heaviside: The Life, Work, and Times of an Electrical Genius of the Victorian Age*, The Johns Hopkins University Press, 2002 (originally published in 1987 by the IEEE Press, with the 2002 edition containing historical updating).

6. Such a speed was achieved, of course, by first punching a paper tape with the message to be sent, and then running the tape at high speed through an automatic transmitter. The mathematical details of Heaviside's work can be found in *Oliver Heaviside* (note 5), or in *Transients for Electrical Engineers* (note 6 in Chapter 1).

7. It was in Heaviside's 1887 paper that the analysis I gave you in chapter 5—resulting in (5.2.6)—first appeared. Cable loading was so contrary to intuition that Heaviside was long declared to be "crazy" by lofty authorities in the British electrical community, a story told in my book, *Oliver Heaviside* (note 5). In the end, when it was realized Heaviside was right, cable loading eventually made a lot of money for some (but not a dime for Heaviside).

8. *Leaked*, because the reason the British could decipher the coded Zimmermann telegram is that her espionage experts had broken the German diplomatic code. If England had just openly handed the deciphered telegram to the Americans, Germany would have realized her code was compromised and so of course have changed it, which the British obviously wanted to avoid. You can read more about this bizarre affair in the classic book by Barbara Tuchman, *The Zimmermann Telegram* (first published in 1958, and updated several times since).

9. After the French mathematical physicist Jean-Marie Duhamel (1797–1872), even though it isn't clear the integral is actually due to him. I think he is better known to mathematicians than to physicists, and I suspect not one-in-a-thousand engineering students today have ever even heard of Duhamel.

10. Just to refresh your memory, if we start with the AP-calculus result for the differential of a product $(d(uv) = u\,dv + v\,du$, where u and v are both functions of τ), we have the amazingly useful $\int_a^b u\,dv = (uv)\Big|_a^b - \int_a^b v\,du$.

11. The tidal analyzer inspired the American physicist Albert Michelson (1852–1931), winner of the 1907 Nobel Prize in physics, to construct (in 1898) a more advanced harmonic analyzer using gears, spiral springs, pinions, and eccentric wheels. With this new design, Michelson—and his co-inventor Samuel Wesley Stratton (1861–1931)—claimed "it would be quite feasible to increase the number of [terms in a Fourier series] to several hundred or even to a thousand with a proportional increase [in accuracy]." The definitive work (with many photographs) on their gadget is Bill Hammack, Steve Kranz, and Bruce Carpenter, *Albert Michaelson's Harmonic Analyzer: A Virtual Tour of a Nineteenth Century Machine That Performs Fourier Analysis*, Artic Noise Books, 2014.

12. See "The Tide Gauge, Tidal Harmonic Analyzer, and Tide Predictor," reprinted in volume 6 of *Kelvin's Mathematical and Physical Papers*, Cambridge University Press, 1911, pp. 272–305. You can find an illustration of the tidal analyzer (it was a pretty amazing gadget!) in Crosbie Smith and M. Norton Wise, *Energy and*

Empire: A Biographical Study of Lord Kelvin, Cambridge University Press 1989, p. 371.

13. A story told in my book, *The Logician and the Engineer*, Princeton University Press, 2013. The logician is the Irish mathematician George Boole (1815–1864), after whom the mathematics of digital computers (Boolean algebra) is named, and the engineer is the American mathematical electrical engineer Claude Shannon (1916–2001), the inventor/discoverer of information theory. Shannon's friend, the English mathematician and computer pioneer Alan Turing (1912–1954), who conceived the stored program concept, also appears in the book.

14. It appears on page 46 in the issue of May 16, 1872. Signed only by "$\frac{dp}{dt}$" its authorship must have been a mystery to most of *Nature*'s readers. The curious signature was an inside-joke between Maxwell and his scientific friends. (Google "dp/dt+maxwell" and learn its significance: even a genius like Maxwell could, now and then, be a goofy guy.)

15. In Oliver Heaviside, "On the Speed of Signalling through Heterogeneous Telegraph Circuits," *Philosophical Magazine*, March 1877, pp. 211–221.

16. The units of T are farads ohms, which is equivalent to seconds. Can you show that? T *must*, of course, have the units of time, since the exponent $\frac{t}{T}$ is dimensionless (do you see why?—Think of what the power series expansion of $e^{-\frac{t}{T}}$ looks like, and then ask yourself what the units of each term must be).

Appendix

1. An integral is defined by a limiting process, and so taking the $\frac{1}{\Delta y}$ inside the integral means we are reversing the order of two limiting processes, something a mathematician would want to justify. Engineers and physicists, however, while aware of the potential risk in doing that, go ahead and do it anyway.

2. The reason I've written a *partial* derivative *inside* the integral, and a *total* derivative *outside* the integral, is that the integrand is a function of two variables (x and y), while the integral itself is a function only of y.

3. The big question that students always ask is, *how do we know to make this definition*? Well, that's the puzzle of human genius, and that's why people who are the first to make such useful definitions generally become famous. In this particular case, however, I have to admit I don't know who first used this trick. The integral was first evaluated (in a different way) in 1774 by Laplace, Fourier's eventual colleague in the French Academy.

4. Reinhold Remmert, *Theory of Complex Functions*, Springer 1991, p. 408.

5. You can find an AP-calculus derivation of Laplace's integral in my *Inside Interesting Integrals*, Springer 2015, pp. 79–82.

Acknowledgments

1. You can read more about this joint work (concerning the mathematics of radio engineering) in the collaborative paper by Shawnee L. McMurwood and James L. Tattersall, "The Mathematical Collaboration of M. L. Cartwright and J. E.

Littlewood," *American Mathematical Monthly*, December 1996, pp. 833–845. Cartwright was Mary Cartwright (1900–1998), who was Mistress of Girton College, Cambridge (Hardy was her thesis advisor when she earned her doctorate in 1930 at Oxford).

2. Anthony Lane, writing in *The New Yorker*, June 18, 2018, p. 64.

INDEX

ALSO BY PAUL J. NAHIN

Paul J. Nahin is professor emeritus of electrical engineering at the University of New Hampshire. He is the author of 21 books on mathematics, physics, and the history of science, published by the university presses of Princeton and Johns Hopkins, and Springer/New York. He received the 2017 Chandler Davis Prize for Excellence in Expository Writing in Mathematics (for his paper "The Mysterious Mr. Graham," *Mathematical Intelligencer*, Spring 2016). He gave the invited 2011 Sampson Lectures in Mathematics at Bates College, Lewiston, Maine.

CPSIA information can be obtained
at www.ICGtesting.com
Printed in the USA
LVHW080826180322
713444LV00009B/16